SEMIOTIC MODELLING OF THE STRUCTURE

John-Tagore Tevet

Research Group of Structure Semiotics
The University Euroacademy
Tallinn
Estonia

&

Honorary Professor
Institute of Mathematics
H-501 Palam Vihar
Gurgaon
India

john.tevet@graphs.ee

www.graphs.ee

www.dharwadker.org/iom

Proceedings of the Institute of Mathematics
Published by Amazon Books

Abstract

Must be stated that the structure as a such, be studied very little, and if so, then either very generally or linked to specific objects. The structure is an integral attribute of all discrete phenomena, their *constructional or organizational side*. Unfortunately, the concept of structure has devalued to a vague adjective of all objects.

In an attempt to explain the essence of structure, we proceed from the fact that the structure is given as a graph and isomorphic graphs have the same structure. However, the structure is something qualitative which with only quantitative graph-theoretical tools is not considered. It has been shown that the essence of structure consists in the relationships between its elements, and it has created a need to look the ways for presenting these relationships.

Here is presented a way for recognition of the structure with exactness up to isomorphism and other structural properties. It is implemented in the form of a semiotic model that enables to explain the essential properties of structure and their transformations.

John-Tagore Tevet
Tallinn, Estonia
November 2014

Contents

1. INITIAL PRINCIPLES

Semiotic modeling of the structure has emerged from the need to bring into order and to explain the general meaning of structure as such. Here is presented a way for recognition of the structure with exactness up to orbits (positions), isomorphism and other structural properties. It is implemented in the form of a semiotic model that enables to explain the essential properties of structure and their transformations.

1.1. Essence of structure

What's the common between such associations or connected sets as a ***system, structure***, as well a ***graph***. All three consist of *elements* and their *relationships*. The system has *many aspects* and depends from the *empirical properties* of the elements and relationships. In the system has essential role its ***function*** and ***structure***.

Structure is a specific characteristic of the associations, an inseparable attribute of all the really existing systemic objects. Structure must be the *constructional* or *organizational side* of the systemic object [20]. Unfortunately, is the concept of structure devalued to fuzzy adjective of any object. Structure exists there where the relations between element pairs are recognizable [39, 40]. These relations are simple recognizable in case of chemical compounds, genetic formations, some networks etc.

Structure constitutes *an abstraction of the system*, its "skeleton", where its elements and relationships are lose its empirical properties but retain their qualitative distinctness in the form of different *positions* in the structure. Structure (Latin word *structura (inner) building*) is defined as a *connection, permanent relationship* or *organizational way* of system's elements [25]. Be argued that to the *model (interpreter, explicate)* of the structure is a ***graph***. All the structural properties are explained on a graph, including the *positions*.

Concepts of *system, structure, position and graph* are easily and pictorially explainable on the ***Rubik's Cube***.

Example 1.1. To this end, let's look at the Rubik's Cube and answer to two questions: 1) What kind of positions have the elements of cube? 2) With turning a layer of the cube changes its structure or system?

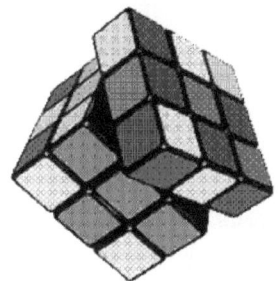

Answer 1. In Rubik's cube has each facet 9 elements, so on all the facets are 6x9 = 54 elements. Each facet has one element in the *middle*, four elements in the *edges* and four elements in the *angles*. Thus, the 6 elements of the cube represent a ***"middle position"***, 24 elements an ***"edge position"*** and 24 elements an ***"angle position"***.

Answer 2. With turning the layers of the cube *changes its system*, because the relationships between its empirical properties of the elements (i.e. colors) changed. However, the **structure does not change**, because the **positions are remained**, i.e. stay to the **invariant**.

The *structure* of Rubik's cube can depict in the form of a **graph** (here can be remark that Rubik's cube as a system has also some aspects – its elements can be also its angles and edges). For present case, each element of this cube has four neighbors: "upper", "lower", "right", "left" and can be presented as a graph, where its 54 vertices divide to **the three positions** – to *"middle"*, *"edge"* and *"angle"* position. As a rule is every structure presentable in the form of a graph and is intimately related with **invariance** and **isomorphism**.

How different the structure from a graph?

Propositions 1.1. On the *difference of the structure and a graph*:
P1.1. *Isomorphic graphs have the same structure*, non-isomorphic different structures, in the other words, *structure is a complete invariant of isomorphic graphs*.
P1.2. If for the presentation of a graph is sufficient to present its adjacency matrix then *for recognition of structure **must the relationships between elements be identified with exactness up to the orbits (positions)**.*

The orbits are in the group theory known as *transitivity domain of automorphisms* or *eqiuvalence classes*, we call it to **positions**. The difference consists only in detection techniques. Identifiers of the element pairs called **binary signs**. An ordered system of binary signs forms the **semiotic model of structure**, which presents the structure with exactness up to positions and isomorphism [28 - 40].

This problem is **heuristic**. To research objects of Heuristics are the *thinking and creative processes, complicated systems and their formalization efforts*. Modern heuristic is connected with the problems of **artificial intelligences**. In case of complicated problems of the graphs can be their treatment with exact mathematical methods impossible. Heuristics arise as soon as there appear an alternative – and the solving almost always subjective. As we see heuristics is unavoidable for IT and other fields.

To one of heuristic methods is **semiotics**. It is a discipline of the *signs and sign systems* that study the phenomena of the *meaning, communication- and interpretation*. The development and implementation of majority the semiotic methods based on the investigation of such systems, which have sufficiently clearly expressed *structure* and sufficiently clear means for *expression of their attributes*. Thereafter, I will try to show that these conditions satisfy the graphs. Semiotic investigations enable formalizing the new objects and in the border areas emerged disciplines, as well as for research of known objects on a *new aspect*.

Sign of a sensually perceptible object (a thing, phenomenon, condition, events) that *represents, sign or describe* from its self different object, its *properties, meaning or sense*. You might say that sign is the *concentrate of an object,* by help which stored, processed and transmitted the information. An object is a sign only in a sure relation with the other signs which partake in the same process and are the same type. In case of a sign is concretized its *meaning* (i.e. indication function, which it represents), and its *sense* (i.e. with a sign associated content of sense). Sign is related with **cognition** and **thinking**. Sign is an essential characteristic of an object. For these reasons is here suitable use the **semiotic modeling**.

Semiotics is characterized with pluralism. The semiotics exist many, in the areas of arts and science. By W. Nöth [21] exist a **semiosphere**, whereto belong from the semiotics of culture to the computer semiotics. One of the first semiotics was *Semiotics of Mathematics* [12]. Semiotics can be related with *computing and artificial intelligence.* Semiotics is *interdisciplinary.* Semiotic modeling can be one of the many kinds of object-oriented semiotics.

To structural signs are the **semiotic invariants** of the graphs [36]. Thus, for structure recognition must the element pairs in the *structure model* identified with exactness up to **binary positions**, i.e. *the positions of element pairs in the structure).*

1.2. Semiotic model of structure

Semiotic model of structure **SM** is an ordered system of identifiers of vertex pairs, i.e. system of binary signs.

Let a vertex pair *ij* can be identified by a specific "relationship" between their in the form of *intersection of neighborhood,* $N_i \cap N_j$, as a partial graph, called **binary graph**. Corresponding **semiotic identification algorithm SIA** find all the binary graphs and determines their *semiotic invariants,* in the form of **basic binary sign** [23].

Semiotic Identification Algoritm (SIA). OPERAND: *List the adjacent vertices L.* ALGORITHM: 1) Fix an element *i* and form its neighborhood N_i, where the elements, connected with *i*, are divided according to distance *d* to entries C_d. 2) Fix an element *j* and form its neighborhood N_j by condition (1). 3) Fix the intersection $N_i \cap N_j$, as a *binary graph* g_{ij}, and fix its invariants in the form of a binary sign $\pm d.n.q._{ij}$. 4) Realize (1) to (3) for each pair *ij*, $i,j \in [1, |V|]$. 5) Obtained preliminary semiotic model. Fix for each vertex (row) *i* its *frequency vector* u_i of pair signs. 6) Decompose the preliminary model **SM** by *frequency vectors* u_i lexicographically to partial models SM_k. 7) In the framework of SM_k decompose the rows and columns by *class vectors* s_i lexicographically to complementary partial models SM_k. 8) Repeat (7) up to complementary decomposing no arise. RESULTS: **a)** *Semiotic model* **SM**; **b)** *The lists of vertices* $\{B_{ij}\}$ *of binary graphs.*

Explanation: The **basic binary sign** is a quadruplet $\pm d.n.q._{ij}$, where $+d$ show collateral- and $-d$ ordinary distance between vertices v_i and v_j, n – number of vertices and q – number of edges, in this binary graph g_{ij}.

It has been suggested the binary sign to *measure (size)* called. Indeed, it has the properties of measure. But yet, this is pointless, because in present case, needed a **description** of the condition of vertex pairs in the structure.

Example 1.2. Graph **G**, its binary signs and **model SM**:

$$A:-2.5.7; \quad B:-2.5.6; \quad C:+2.3.3; \quad D:+2.5.7; \quad E:+3.6.10.$$

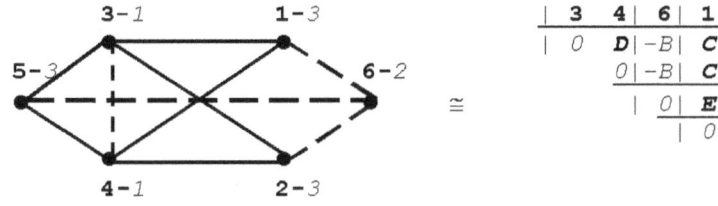

1	1	2	3	3	3		u_i	k	s_i
3	4	6	1	2	5	*i*	ABCDE		123
0	D	-B	C	C	C	3	01310	1	103
	0	-B	C	C	C	4	01310	1	103
		0	E	E	E	6	02003	2	003
			0	-A	-A	1	20201	3	210
				0	-A	2	20201	3	210
					0	5	20201	3	210

Explanation: The vertices have **three positions** and vertex pairs are in **five positions**, including *three "edge" positions* (full line **C**, broken line **E** and broken line **D**) and *two "non-edge" positions*.

For studying the structural properties is necessary *knowing the basic binary signs*, understand their meaning, ability *to read the model* **SM** *as a text of graph structure*.

Propositions 1.2. The **basic binary signs** $\pm d.n.q._{ij}$ describes the conditions of vertex pairs in the structure:

P1.2.1. Binary sign in the form $-d.n.q._{ij}= -\infty.2.0$ signify a **disconnected vertex pair.**

P1.2.2. Binary sign in the form $-d.n.q._{ij}$ signify that the vertex pair forms a **simple path** or their *assemblage* with length d and we call it **path sign.**
For example: $-2.3.2$ shows a 2-path; $-3.4.3$ shows a 3-path; $-7.8.7$ shows a 7-path, etc.

P1.2.3. Binary sign with greatest absolute value $max|-d|$ show the **diameter** of the structure.

P1.2.4. Binary sign $+d.n.q._{ij}=+1.2.1$ signify that the vertex pair forms a **link of a branch** and we call it **branch sign.**

P1.2.5. Binary sign $+d.n.q._{ij}$, where $+d\geq2$, signify that the vertex pair belong to **girth** or their *assemblage* with length $d+1$ and it is called **girth sign.**
For example, girth signs are, $+3.4.4$ for 4-girth; $+4.5.5$ for 5-girth; $\ldots+7.8.8$ for 8-girth etc. If the length $+d$ no correspond with the numbers of vertices n and edges q, then is touch with any mutually intercrossed $(d+1)$-girths.

P1.2.6. Binary sign in the form $(+d=2).n.(q=n(n-1):2)$ signify belonging to a **clique** and we call it **clique sign.**
For example: $+1.2.1$ show a 2-clique; $+2.3.3 - 3$-clique; $+2.4.6 - 4$-clique; $+2.5.10 - 5$-clique; $+2.6.15 - 6$-clique; $\ldots$, $+2.13.78 - 13$-clique etc. Clique sign is a **complete invariant** of the clique, i.e. to clique sign correspond only a clique.

It is useful treat also the **accompanying graphs** of a graph, such as *complement, pair graphs, sign graphs* and *adjacent graphs*.

The structure be studied (investigates) *in an integrated way*, in conjunction with its *complement*.

Example 1.3. The **complement** $\rceil G$ of graph G (on example 1.2) and its **model SM**:

$$A:-2.3.2; \quad B:-0.2.0; \quad C:+1.2.1; \quad D:+2.3.3.$$

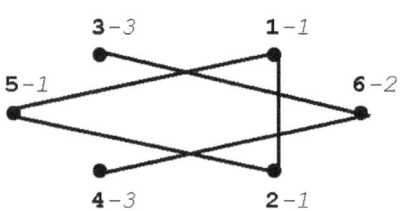

	1	1	1	2	3	3			$\mathbf{u_i}$	$\mathbf{k}$	
	1	2	5	6	3	4		$\mathbf{i}$	ABCD		123
	0	D	D	$-B$	$-B$	$-B$		1	0302	1	200
		0	D	$-B$	$-B$	$-B$		2	0302	1	200
			0	$-B$	$-B$	$-B$		5	0302	1	200
				0	C	C		6	0320	2	002
					0	$-A$		3	1310	3	010
						0		4	1310	3	010

Explanations: **a)** The **vertex and binary positions** in G and its *complement* $\rceil G$ *remain always*. **b)** But if G has *three edge and two non-edge positions* then in complement $\rceil G$ is it in *contrary*.

Binary graph be characterize the condition of a vertex pair, *binary sign* is only its invariant. In some cases, it is necessary to open the *structure model of binary graph* for adjustment of corresponding binary sign.

Example 1.4. **Binary graph $g_{3.4}$** of the vertex pair 3-4 *(D)* of *G* and its *model* SM:

A:-2.4.5; B:-0.2.0; C:+2.3.3; E:+2.5.7.

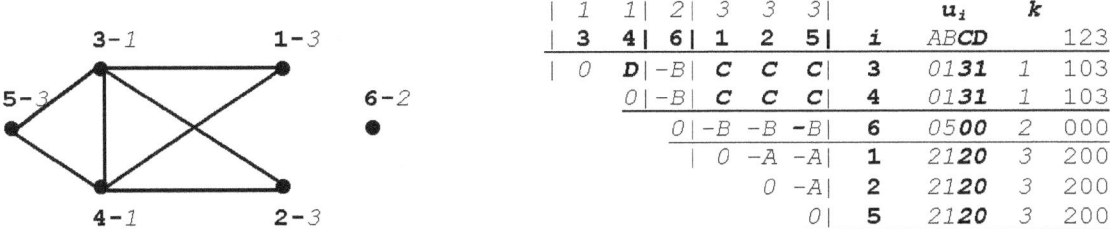

	1	1	2	3	3	3				
	3	4	6	1	2	5	i	ABCD (u_i)	k	123
	0	D	-B	C	C	C	3	0131	1	103
		0	-B	C	C	C	4	0131	1	103
			0	-B	-B	-B	6	0500	2	000
				0	-A	-A	1	2120	3	200
					0	-A	2	2120	3	200
						0	5	2120	3	200

Explanation: The *elements*- and *binary positions* of *G* and its binary graph $g_{3.4}$ *remain* but this is not regular.

Sign graph is one of the key attributes of structure, by it can *adjust the pair signs* (Prop. 1.3) and investigate the *structural properties*. In some cases it may be turn out to the *position structure* (Prop. 2.9).

Example 1.5. **Sign graph $G_{p=+E}$** of by sign *E: +3.6.10* of *G* and its *model* SM:

A: -2.3.2; B: -0.2.0; C: +1.2.1.

	1	2	2	3	3	3				
	6	3	4	1	2	5	i	ABC (u_i)	k	123
	0	-B	-B	C	C	C	6	023	1	003
		0	-B	-B	-B	-B	3	050	2	000
			0	-B	-B	-B	4	050	2	000
				0	-A	-A	1	221	3	100
					0	-A	2	221	3	100
						0	5	221	3	100

Explanations: **a)** Sign graphs $G_{p=+C}$ $G_{p=+D}$ and $G_{p=+E}$ of *G* are its *partial graphs*, whereby $G_{p=+C}$ cover *G* completely and $G_{p=+D}$ only vertex pair 3-4. **b)** Sign graphs $G_{p=-A}$ and $G_{p=-B}$ are partial graphs of complement $\rceil G$. **c)** Also here the *positions remain* but this is not regular.

*

Structural models are useful also for investigation of ***natural objects***. For example, structural model of the *chemical compound* is a detailed presentation of the classical structural formula, i.e. of a *graph* that represents this formula. Structural model recognizes all the relationships between elements and their positions.

Example 1.6. Structural formula (i.e. graph) of *isobutan* C_4H_{10}, its binary signs and structural model:

$$A:-4.5.4; \quad B:-3.4.3; \quad C:-2.3.2;$$
$$D:+1.2.1.$$

2	1	3	4	11	5	6	7	8	9	10	12	13	14	a	i	u_i	k	s_i
C	C	C	C	H	H	H	H	H	H	H	H	H	H			ABCD		1234
0	D	D	D	D	-C	-C	-C	-C	-C	-C	-C	-C	-C	C	2	0094	1	0310
	0	-C	-C	-C	D	D	D	-B	-B	-B	-B	-B	-B	C	1	0634	2	1003
		0	-C	-C	-B	-B	-B	D	D	D	-B	-B	-B	C	3	0634	2	1003
			0	-C	-B	-B	-B	-B	-B	-B	D	D	D	C	4	0634	2	1003
				0	-B	-B	-B	-B	-B	-B	-B	-B	-B	H	11	0931	3	1000
					0	-C	-C	-A	-A	-A	-A	-A	-A	H	5	6331	4	0100
						0	-C	-A	-A	-A	-A	-A	-A	H	6	6331	4	0100
							0	-A	-A	-A	-A	-A	-A	H	7	6331	4	0100
								0	-C	-C	-A	-A	-A	H	8	6331	4	0100
									0	-C	-A	-A	-A	H	9	6331	4	0100
										0	-A	-A	-A	H	10	6331	4	0100
											0	-C	-C	H	12	6331	4	0100
												0	-C	H	13	6331	4	0100
													0	H	14	6331	4	0100

Explanation: Decomposition of the elements C and H to four positions is true.

This is a so-called systemic approach to the study of chemical compounds where different chemical elements (atoms) as a rule, are divided into different *positions* and can be treated as *subsystems*. In case of more complex compounds, however, may also the same kind of elements (atoms) belong to different positions (for example, ethanol, butane, propane, etc.). The main idea of using the models consists in treatment of the whol,e on the basis of positions and the relationships between them. Structural models open up the possibility for additional investigation of chemical compounds. The models of some polymers and organic matters tend to be very large [40]. Here be limited with a moderate.

Structural model allows to recognize: 1) *the structure of graph*, where an essential meaning have the *binary- and vertex- (elements-) positions* as well as the *sign- and position structures*; 2) *equivalence of the structures* and *isomorphism of the corresponding graphs;* 3) the elementary *structural transformations* in the form of *adjacent structures.*

Sign structure is composed of the element pairs that have the same basic binary signs. **Position structure** consists of element pairs that belong to the same position (Def. 2.3). **Adjacent structure** is a greatest substructure or smallest superstructure of a structure which is obtained with removing or adding a connection (edge) in a binary position (Def. 4.1). This is related with the **reconstruction problem** (Ulam Conjecture) and opens a way to formation of the **systems of adjacent structures** i.e. *systems of non-isomorphic graphs with **n** vertices*.

1.3. Adjustment and simplification of the model

In common case are the structures recognizable on the level of basic binary signs, but in case of some symmetric graphs is necessary to use the **adjusted binary signs**. In the case of solving some practical tasks is suitable the binary signs **to simplify**.

Adjustment

It is obvious that basic binary signs in the form *d.n.q.* does not always become to complete identifier of vertex pairs. To exact ascertaining the structure of *some large transitive and symmetric* graphs is needful to **adjust the binary signs** or to **deep-identify** [37]. For this be exist any possibilities. Also here assist the **binary-** and **sign graphs**.

Propositions 1.3. Possibilities for *adjust identifications*:

P1.3.1.   Using complementary pair signs $d.n.q._{ij}{}^{m}$ of the **high degree m binary graphs $g_{ij}{}^{m}$**. We call it **high identification**.

For example, second degree binary graph $g_{ij}{}^{m=2}$ is this, which remain between vertices i and j of G after removing the preliminary binary graph g_{ij}, i.e. $g_{ij}{}^{m=2} = G \setminus [g_{ij} \setminus (v_i \& v_j)]$.

P1.3.2.   Using complementary binary signs of the *local structure model* SM_{ij} of first or high degree **binary graphs g_{ij}**. We call it **local identification**.

P1.3.3.   Using complementary binary signs of the **structure model** SM_p of a **sign graph G_p**. Such deep identification mode we call **sign graph identification**.

P1.3.4.   Using the exponentation (involution) of adjacency matrices E to a certain degree E^n increases *the values of its elements* as well as *the number of different values*. Increase takes place up to a degree n, after which it stops. We call it **involution identification**.

It turned out that these values of involution identify the element pairs with exactness up to the binary positions (and on their basis the positions of the elements) [33]. The basic binary signs do not lose their meaning. They remain anyway characterize the elements and their relationships between these.

Involution Identification Algorithm (IIA). OPERAND: *List of adjacent elements L.* ALGORITHM: 1) Form the adjacency matrix E. 2) Multiple it with itself $E \times E \times E \times \ldots = E^n$ and fix in case of each degree n the number p of obtained differences among *productive binary signs e^n_{ij}*, which as rule enlarge. 3) If p more no enlarge, then to stop the multiplication and to fix the last product E^n, corresponding p and successive E^{n+1}. 4) Obtained matrix products E^n and E^{n+1}, as certain types "sign matrices". 5) If necessary, to specify the basic binary signs $\pm d.n.q._{ij}$ with the productive, $\pm d.n.q.e^n._{ij}$. RESULT: An adjusted model SM^*.

Explanations: **a)** The maximum number p can be greater than the number of differences among the basic binary signs and in some cases to *adjust* these. **b)** Adjustably identified binary sign, for example $\pm d.n.q.e^n._{ij}$ we call to **adjusted sign** and corresponding model to **adjusted model** SM^* (example 3.6).

Example 1.7. Result of involution identification **IIA**: The product $E \times E \times E = E^3$ of adjacency matrices and adequacy of basic and productive binary signs of a transitive graph:

Basic binary signs		0	-2.6.11	+2.5.8	+2.4.5	+2.4.6
Productive binary signs $e_{ij}{}^3$		12	13	16	18	19

1	2	3	4	5	6	7	8	=i	u_i=12345	k
12	18	16	13	19	13	16	18	1	12221	1
18	12	18	16	13	19	13	16	2	12221	1
16	18	12	18	16	13	19	13	3	12221	1
13	16	18	12	18	16	13	19	4	12221	1
19	13	16	18	12	18	16	13	5	12221	1
13	19	13	16	18	12	18	16	6	12221	1
16	13	19	13	16	18	12	18	7	12221	1
18	16	13	19	13	16	18	12	8	12221	1

Explanations: **a)** In present case is the complete identify attainable on the level of the third degree e_{ij}^3. **b)** Productive binary signs no contain direct structural data.

In principle the structure's model could be based only on the productive binary signs, if would be know what these mean. It is only said that the elements of E^n characterize the longest paths (chains) between structural elements. Unfortunately, it is questionable because the values exist also on the main diagonal, at the same time as in elsewhere the values are also at times turn out to be zero. These no distinguish the adjacent and non-adjacent elements but are well suited to refine the basic binary signs.

Apparently no one has interested in the question: *why detect the elements of obtained matrix E^n the binary positions?* Already in 1976 was drawn attention to the too one-sided approach to graphs which hinders the development of graph theory [18].

The *basic binary signs* not lose its meaning, these characterizes the relationships between vertices, the belonging of vertex pairs to (assemblage of) paths or girths with corresponding size etc. These are needed for characterizing of the structure as a whole. In case of some strongly regular graphs where need use the *local identification* method P1.3.2.

Simplification

We had afore shown rather symmetric graphs. For all the graphs generally be valid follow proposition.

Proposition 1.4. Almost all the graphs are *0-symmetric, connected* and with *diameter 2.*

This means, that each vertex and vertex pair constitute a single position in structure. Be lacking any symmetry, the number of pair signs is very large. For solving applicative tasks is necessity to find some *"similarity"* between elements.

Since real **communication networks** are very large. Imagine here one a peculiar companionship **Z** consisting of *Adolf, Birgit, Charles, Diana, Erik, Frieda, George, Helen, Ingvar and Jane*. They are mutually agreed that everyone communicates with the five, known to us, parlor companions. The latter circumstance had required of coordination, and someone had to do it.

This situation show a *five-degree-regular structure* where all the members seem to be in "equal position".

Example 1.8. To present this situation make a corresponding model **Z**:

A:-2.6.10; B:-2.6.9; C:-2.5.8; D:-2.5.7; E:-2.5.6; F:-2.4.5; G:-2.4.4;
H:+2.3.3; I:+2.4.5; J:+2.5.7; K:+3.10.25.

1	2	3	4	5	6	7	8	9	10	name	u_i ABCDEFGHIJK	k
F	A	D	H	C	B	I	J	E	G			
0	-G	I	J	-D	-F	I	-E	I	H	Frieda	00011111310	1
	0	H	-G	J	I	-D	I	-D	I	Adolf	00020021310	2
		0	-C	I	-D	H	-E	I	-D	Diana	00121002300	3
			0	-E	H	I	-B	H	H	Helen	01101013110	4
				0	H	-G	-A	H	H	Charles	10011013110	5
					0	I	I	-E	-A	Birgit	10011102300	6
						0	H	-A	-D	Ingvar	10020012300	7
							0	K	H	Jane	11002002201	8
								0	-B	Erik	11011002201	9
									0	George	11020004100	10

The structure Z is *0-symmetric*, there do not "equality", each member has its own private position. *Different position* means different connectivity, "relationships" with other members. Between ten members exists 11 different relationships, which is characterized by the binary signs (see frequency vectors u_i). The problem lies here in the *grouping of strictly differentiated members*. This fact leads us back to the *sign structures* GS_p. In selection of the sign must be proceeds from:

1) Selected sign must be exists *in case of each structural element*.
2) To keep in mind the *meaning of sign*, because the sign structure be formed on the aspect of sign.

In principle is the companionship decomposable to the eleven inseparable component sign structures GS_p, and gives different groupings. This is inappropriate, and useful to go the other way.

Let to it is the rearranging the members by their "direct communication signs" *HIJK* of *u*-vectors.

Example 1.9. Rearranged by *HIJK* model *Z*:

1	2	4	5	3	6	7	8	9	10	name	k	HIJK	R
F	A	H	C	D	B	I	J	E	G				
0	-G	J	-D	I	-F	I	-E	I	H	Frieda	1	1310	1
	0	-G	J	H	I	-D	I	-D	I	Adolf	2	1310	1
		0	-E	-C	H	I	-B	H	H	Helen	4	3110	2
			0	I	H	-G	-A	H	H	Charles	5	3110	2
				0	-D	H	-E	I	-D	Diana	3	2300	3
					0	I	I	-E	-A	Birgit	6	2300	3
						0	H	-A	-D	Ingvar	7	2300	3
							0	K	H	Jane	8	2201	4
								0	-B	Erik	9	2201	4
									0	George	10	4100	5

The resulting grouping corresponds to the requirement of "direct communication signs", where the *ten positions k* reduces to *five groups*, with the members:

R_1= (*Frieda, Adolf*), R_2= (*Helen, Charles*), R_3= (*Diana, Birgit, Ingvar*),
R_4= (*Jane, Erik*) and R_5= (*George*).

For finding the "similarity" of members can be use also *approximate or rounded-off* binary signs.

Example 1.10. Using the rounded-off binary signs:

Rounding-off: a= [A:-2.6.10; B:-2.6.9], b= [C:-2.5.8; D:-2.5.7; E:-2.5.6],
c= [F:-2.4.5; G:-2.4.4], d= [H: +2.3.3; I:+2.4.5; J: +2.5.7], e= (K: +3.10.25).
Rounded binary signs: a:(A, B) $\approx$ –2.6, b:(C, D, E) $\approx$ –2.5, c:(F, G) $\approx$ –2.4, d:(H, I, J) $\approx$ +2 ja e: K $\approx$ +3.

1		2	3		4		5	name	AB	CDE	FG	HIJ	K	k	abcde	k*
F	A	D	H	C	B	I	J	E	G								
0	-G	I	J	-D	-F	I	-E	I	H	Frieda	00	011	11	131	0	1	02250	1
	0	H	-G	J	I	-D	I	-D	I	Adolf	00	020	02	131	0	2	02250	1
		0	-C	I	-D	H	-E	I	-D	Diana	00	121	00	230	0	3	03050	2
			0	-E	H	I	-B	H	H	Helen	01	101	01	311	0	4	12150	3
				0	H	-G	-A	H	H	Charles	10	011	01	311	0	5	12150	3
					0	I	I	-E	-A	Birgit	10	011	10	230	0	6	12150	3
						0	H	-A	-D	Ingvar	10	020	01	230	0	7	12150	3
							0	K	H	Jane	11	002	00	220	1	8	22041	4
								0	-B	Erik	11	011	00	220	1	9	22041	4
									0	George	11	020	00	410	0	10	22050	5

The resulting grouping by rounded-off binary signs:

$$k^*_1 = (Frieda, Adolf), \quad k^*_2 = (Diana), \quad k^*_3 = (Helen, Charles, Birgit, Ingvar),$$
$$k^*_4 = (Jane, Erik) \text{ and } k^*_5 = (George).$$

We can see that there exist coincidences between the results of "direct communication signs" and rounding-off. The first way shall be considered as more distinct and therefore more reliable. The "rounding" of binary signs may prove to be quite arbitrary. Here can remark a *specific role of Jane and Erik* in this companionship, to their relationship $K: +3.10.25$ includes all members and relationships, and they may be *coordinators*.

Such *0*-symmetric structures can be treats, investigate, and elements grouped in several ways:
1) By investigation of the selected sign structures GS_p.
2) By investigate on the basis of some selected binary signs formed the so-called complex sign structures.
3) By reordering the structural model by the given binary signs (example 1.9).
4) For reducing the positions to use the connected or "rounded" binary signs (example 1.10).

All of this requires a good knowledge of the subject and suitable choices the aspects for the investigation.

2. PROPERTIES OF STRUCTURE

The main structural properties are connected with *regularity* and *positions (symmetry)* of structure. Structural properties are read out from basic binary signs in the structure model **SM** [35].

2.1. Diversity of structural regularity

Propositions 2.1. On the ***regularities:***

P2.1.1. Graph, where the numbers of partial binary(+)signs *+d* in all the rows *i* of **SM** are equal is *(degree)-regular.*

P2.1.2. Graph, where the partial signs *–d* of all the binary(–)signs *–dnq* in **SM** equal is *d-distance-regular*.

On example 2.1 showed Petersen graph with its pair(–)sign *–2.3.3* is *2-distance-regular*.

P2.1.3. Graph, where the partial signs *+d* of all the binary(+)signs *+dnq* in **SM** are equal is *(d+1)-girth-regular.*

For example, Petersen graph with its pair(+)sign *+4.10.15* is *5-girth-regular* (example 2.1). In girth-regular graph belong all the *n* vertices the same number times to girth with length *n–a*.

P2.1.4. Graph, where the numbers of clique signs *+(d=2).n.(q=n(n-□1):2)* in all the rows *i* of **SM** are equal is *n-clique-regular.*

For example, the complement of Petersen is *4-clique-regular* (example 2.1). If *a transitive* graph self not clique, then there can not exists a single clique, it can be only clique regular. In clique-regular graph belong all the *n* vertices the same number times to clique with power *n–b*.

P2.1.5. Graph said ***strongly regular*** with parameters *(k,a,b)* if it is a *k-degree-regular* incomplete connected graph such that any two adjacent vertices have exactly *a≥0* common neighbors and any two non-adjacent vertices have *b≥1* common neighbors.

For example, Petersen graph is *strongly regular*. Its strong regularity accrues from its bisymmetry.

Example 2.1. Petersen graph ***Pet***, the binary signs and structure model for Petersen graph and its complement ***PetC*** (the numbering starts here from the upper element and goes clockwise):

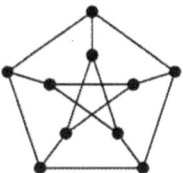

A:-2.3.2; B:+4.10.15.   A:-2.6.12; B:+2.5.8.

1	1	1	1	1	1	1	1	1	1	u_i		
1	2	3	4	5	6	7	8	9	10	*i*	AB	*k*
0	B	-A	-A	B	B	-A	-A	-A	-A	1	63	1
	0	B	-A	-A	-A	B	-A	-A	-A	2	63	1
		0	B	-A	-A	-A	B	-A	-A	3	63	1
			0	B	-A	-A	-A	B	-A	4	63	1
				0	-A	-A	-A	-A	B	5	63	1
					0	-A	B	B	-A	6	63	1
						0	-A	B	B	7	63	1
							0	-A	B	8	63	1
								0	-A	9	63	1
									0	10	63	1

1	1	1	1	1	1	1	1	1	1	u_i
1	2	3	4	5	6	7	8	9	10	AB
0	-A	B	B	-A	-A	B	B	B	B	36
	0	-A	B	B	B	-A	B	B	B	36
		0	-A	B	B	B	-A	B	B	36
			0	-A	B	B	B	-A	B	36
				0	B	B	B	B	-A	36
					0	B	-A	-A	B	36
						0	B	-A	-A	36
							0	B	-A	36
								0	B	36
									0	36

Explanations to show that it is possible to read out from the structure model:
- a) Petersen graph *Pet* is *3-degree-*, *2-distance-* and *5-girth-regular*.
- b) Binary sign *+4.10.15* means, that the element pair belongs to an assemblage of 5-girths, which consists of 10 elements and 15 connections (edges) – it is the *complete invariant* of Petersen graph, such sign do not have other structures.
- c) *Pet* is *5-girth-regular*, there exist twelve *5-girths*, in present case: (1): 1-2-3-4-5-1, (2): 6-8-10-7-9-6, (3): 1-2-3-8-6-1, (4): 1-2-7-10-5-1, (5): 1-5-4-9-6-1, (6): 2-3-4-9-7-2, (7): 3-4-5-10-8-3, (8): 1-2-7-9-6-1, (9): 1-5-10-8-6-1, (10): 2-3-8-10-7-2, (11): 3-4-9-6-8-3, and (12): 4-5-10-7-9-4. Each element belong to six girths, each edge belongs to four girths.
- d) The *complement* of Petersen graph *PetC* is *4-clique-regular*. Explicit clique sign do not exist, but *binary graph* of binary sign *+2.5.8* contains the *4-clique*. For example, the local structure model of binary graph with sign *+2.5.8* for elements 1 and 3 contains the signs of *4-clique*, *+2.4.6*, that shows the existence of *4-clique* 1,3,9,10:

```
-A: -2.4.5;  B: +2.3.3;  C: +2.4.6;  D: +2.5.8.
| 1   3| 9 10| 7|     i   ABCD   k    123
| 0   D| C  C| B|     1   0121   1    121
     0| C  C| B|     3   0121   1    121
       | 0  C|-A|     9   1030   2    210
          0|-A|    10   1030   2    210
             0|     7   2200   3    200
```

- e) And so exists in the complement five intersected *4-cliques*, in present case with elements: (1): 1,3,9,10; (2): 2,4,6,10; (3): 1,4,7,8; (4): 2,5,8,9; and (5): 3,5,6,7. Each element belongs to two cliques, each edge belongs to one clique.

Ideology of almost all the clique algorithms is oriented to recognition only a single maximum clique.

Proposition 2.2. A *transitive,* i.e. vertex symmetric graph is *girth-* or/and *clique-regular*.

Example 2.2. Structure models dodecahedron *Dod* and its complement *DodC*:

```
-A=-5.20.30;  -B=-4.8.9;  -C=-3.4.3;  -D=-2.3.2;  +E=+4.8.9.

 1  2  3  4  5  6  7  8  9 10 11 12 13 14 15 16 17 18 19 20|   i   ABCDE   k
 0  E -D -C -B -C -D  E -D -C -B -A -B -C -D -D -C -C -D  E|   1   13663   1
    0  E -D -C -C -D -D -C -B -A -B -C -D -E -D -C -B -C -D|   2   13663   1
       0  E -D -D  E -D -C -C -B -C -C -D -D -C -B -A -B -C|   3   13663   1
          0  E -D -D -C -B -C -C -D -D  E -D -C -C -B -A -B|   4   13663   1
             0  E -D -C -C -D -D  E -D -D -C -B -C -C -B -A|   5   13663   1
                0  E -D -D  E -D -D -C -C -B -A -B -C -C -B|   6   13663   1
                   0  E -D -D -C -C -B -C -C -B -A -B -C -C|   7   13663   1
                      0  E -D -C -B -A -B -C -C -B -C -D -D|   8   13663   1
                         0  E -D -C -B -A -B -C -C -D  E -D|   9   13663   1
                            0  E -D -C -B -A -B -C -D -D -C|  10   13663   1
                               0  E -D -C -B -C -D  E -D -C|  11   13663   1
                                  0  E -D -C -C -D -D -C -B|  12   13663   1
                                     0  E -D -D  E -D -C -C|  13   13663   1
                                        0  E -D -D -C -B -C|  14   13663   1
                                           0  E -D -C -C -D|  15   13663   1
                                              0  E -D -D  E|  16   13663   1
                                                 0  E -D -D|  17   13663   1
                                                    0  E -D|  18   13663   1
                                                       0  E|  19   13663   1
                                                          0|  20   13663   1
```

Explanation: *Dod* is *3-degree-* and *5-girth regular*.

$$-A=-2.16.102; \quad +B=+2.14.78; \quad +C=+2.14.79; \quad +D=+2.15.89.$$

```
 1  2  3  4  5  6  7  8  9 10 11 12 13 14 15 16 17 18 19 20|   i   ABCCD   k
 0 -A  D  B C1  B  D -A  D  B C1 C2 C1  B  D  D  B  B  D -A|   1   36316   1
    0 -A  D  B  B  D  D  B C1 C2 C1  B  D -A  D  B C1  B  D|   2   36316   1
       0 -A  D  D -A  D  B  B C1  B  B  D  D  B C1 C2 C1  B|   3   36316   1
          0 -A  D  D  B C1  B  B  D  D -A  D  B  B C1 C2 C1|   4   36316   1
             0 -A  D  B  B  D  D -A  D  D  B C1  B  B C1 C2|   5   36316   1
                0 -A  D  D -A  D  D  B  B C1 C2 C1  B  B C1|   6   36316   1
                   0 -A  D  D  B  B C1  B  B C1 C2 C1  B  B|   7   36316   1
                      0 -A  D  B C1 C2 C1  B  B C1  B  D  D|   8   36316   1
                         0 -A  D  B C1 C2 C1  B  B  D -A  D|   9   36316   1
                            0 -A  D  B C1 C2 C1  B  D  D  B|  10   36316   1
                               0 -A  D  B C1  B  D -A  D  B|  11   36316   1
                                  0 -A  D  B  B  D  D  B C1|  12   36316   1
                                     0 -A  D  D -A  D  B  B|  13   36316   1
                                        0 -A  D  D  B C1  B|  14   36316   1
                                           0 -A  D  B  B  D|  15   36316   1
                                              0 -A  D  D -A|  16   36316   1
                                                 0 -A  D  D|  17   36316   1
                                                    0 -A  D|  18   36316   1
                                                       0 -A|  19   36316   1
                                                          0|  20   36316   1
```

Explanation: The complement **DodC** is **2-distance- and 16-degree regular.**

Is the complement **DodC** of **5-girth-regular** dodecahedra **Dod** clique-regular?

Example 2.3. In complement **DodC** the explicit clique signs no exist, but in the processing the binary graphs g_{ij}, for example with signs **+B=+2.14.78**, obtained local structure models $SM_{1.4}$, $SM_{5.9}$, $SM_{3.16}$, $SM_{6.13}$ and $SM_{5.8}$, contain *8-clique signs* **+2.8.28**. On the ground of such local structure models can be to recognize all the "hidden" **partial 8-cliques** of **DodC**:

i=	1	2	3	4	5	6	7	8	9	10	11	12	13	14	15	16	17	18	19	20
I	●			●			●			●		●			●		●		●	
II		●			●		●		●		●			●			●			●
III	●		●			●			●			●		●		●		●		
IV		●		●		●		●			●		●			●			●	
V			●		●			●		●			●		●			●		●

Explanation: Thus, the complement **DodC** is **8-clique-regular**, where all five partial cliques are **intercrossed**, and where all the 10 intercrossing edges belong to binary orbit **C2**.

i-j=	1-12	2-11	3-18	4-19	5-20	6-16	7-17	8-13	9-14	10-15
Partial clique	I	II	III	I	II	III	I	IV	II	I
Partial clique	III	IV	V	IV	V	IV	II	V	III	V

Form known graphs are *clique regular* also complements of Heawood's, Coxeter's, Folkman's graphs. Their originals are bipartite and by all the nature laws represent the complements of such parts self-evidently cliques.

Proposition 2.3. Complement of a **m-partite** graph in case of equal **n** parts is **n-clique regular**, with the number **m** of non-intercrossed **n-cliques.**

Proposition 2.4. Partial cliques of a clique regular graph can be *disconnected partial, mutually connected* or *intercrossed.*

Intercrossing can be exists on the aspect of vertices and edges. For example: cliques of *PetC* intercrossed by vertices, of *DodC* by edges.

Example 2.4, Heawood graph *Hea* (the numbering starts from the upper element and goes clockwise) the structure model of *Hea* and its complement *HeaC*:

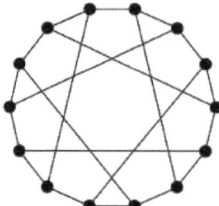

A:-3.8.9; B:-2.3.2; C:+5.14.21.

1	2	3	4	5	6	7	8	9	10	11	12	13	14	i	ABC	k	deg
0	+C	B	A	B	+C	B	A	B	A	B	A	B	+C	1	463	1	3
	0	+C	B	A	B	A	B	A	B	+C	B	A	B	2	463	1	3
		0	+C	B	A	B	+C	B	A	B	A	B	A	3	463	1	3
			0	+C	B	A	B	A	B	A	B	+C	B	4	463	1	3
				0	+C	B	A	B	+C	B	A	B	A	5	463	1	3
					0	+C	B	A	B	A	B	A	B	6	463	1	3
						0	+C	B	A	B	+C	B	A	7	463	1	3
							0	+C	B	A	B	A	B	8	463	1	3
								0	+C	B	A	B	+C	9	463	1	3
									0	+C	B	A	B	10	463	1	3
										0	+C	B	A	11	463	1	3
											0	+C	B	12	463	1	3
												0	+C	13	463	1	3
													0	14	463	1	3

A:-2.10.36; B:+2.8.22; C:+2.9.30.

1	2	3	4	5	6	7	8	9	10	11	12	13	14	i	ABC	k	deg
0	A	+C	+B	+C	A	+C	+B	+C	+B	+C	+B	+C	A	1	346	1	10
	0	A	+C	+B	+C	+B	+C	+B	+C	A	+C	+B	+C	2	346	1	10
		0	A	+C	+B	+C	A	+C	+B	+C	+B	+C	+B	3	346	1	10
			0	A	+C	+B	+C	+B	+C	+B	A	+C	A	4	346	1	10
				0	-A	+C	+B	+C	A	+C	+B	+C	+B	5	346	1	10
					0	A	+C	+B	+C	+B	+C	+B	+C	6	346	1	10
						0	A	+C	+B	+C	A	+C	+B	7	346	1	10
							0	A	+C	+B	+C	+B	+C	8	346	1	10
								0	-A	+C	+B	+C	A	9	346	1	10
									0	A	+C	+B	+C	10	346	1	10
										0	A	+C	+B	11	346	1	10
											0	A	+C	12	346	1	10
												0	A	13	346	1	10
													0	14	346	1	10

Explanations:
a) Graph *Hea* is *3-degree-* and *6-girth-regular*. From 6-girth regularity concluded its *bipartite*, where its parts in present case divide to vertices with even numbers and vertices with odd numbers.
b) *Hea* is *structurally unique*, the pair sign *+5.14.21* signify that its 14 vertices form 21 adjacent pairs that belong to *6-girths* and form a *complete invariant* of this graph.

c) From bipartite *Hea* conclude that its complement *HeaC* consist of two **mutually connected 7-cliques,** it is **7-clique regular**, where the cliques correspond to the parts of *Hea. HeaC* is also **2-distance-** and **10-degree regular.**

2.2. Attributes of symmetry – positions

One of the key features of structure is **symmetry.** Symmetry is a structural characteristic that be expressed as a *recurrence* (in space or time) the similar parts (elements) of an object [20]. In this sense, represents symmetry an **equivalence class**, which consists of "similar" elements, and they form a **position** in the structure. Position characterized by *isomorphism of its accompanying graphs*.

However, it is a widespread understanding of *symmetry* characteristic, where the parts (elements) take similar then, if these are located from a central point, or an axis on the same distance [25]. Such widespread is in mathematics defined as: a) the shape feature "transform to itself" (e.g. isometric); b) feature of binary relation $xRy \leftrightarrow yRx$. Directed graphs called such a link (edge) as well as symmetrical.

Propositions 2.5. The relationships between **positions** and corresponding **subgraphs.**

P2.5.1.   If vertices v_i , v_j , ... have in graph *G the same position* ΩV_k then corresponding *subgraphs* $(G_i=G\backslash v_i) \cong (G_j=G\backslash v_j) \cong....$ *are isomorphic.*

P2.5.2.   Edges e_{ij}, e_{i*j*}, ... have in graph *G the same binary position* ΩR_n then corresponding *subgraphs* $(G_{ij}=G\backslash e_{ij}) \cong (G_{i*j*}=G\backslash e_{i*j*}) \cong....$ *are isomorphic* .

General concept of symmetry is defined in mathematics by **automorphism** α, as a **substitution which retains the structure**. It is treated also as an **inner-** or **local isomorphism** *(isomorphism with itself)*. Substitution has also been associated with renumbering of elements. In fact, none renumbering does not change the structure, it changes only the *removal, addition or relocation of a edge*. The automorphisms form the automorphism group of graph *AutG* where its *transitivity domains to* **orbits** Ω called. An orbit is practically the same *equivalence class* what coincide whit previous *position class*. In case of *AutG* be interested primarily on vertex orbits.

In the *frame of vertex positions* have also the *vertex pairs own positions* or orbits. Here is suitable issue from **binary positions** or **-orbits.** From now on we stay by term **position**.

Propositions 2.6. The relationships between **automorphisms, local isomorphisms, transitivity domains, binary positions** and **binary signs**:

P2.6.1.   As an *automorphism* α or permutation that retain the structure be expressed in the form of a *local isomorphism* $G^{adj}_{ij} \cong G^{adj}_{i*j*}$ then constitute *transitivity domain of automorphisms* or **binary position** ΩR_n, i.e. an *isomorphism class of adjacent graphs* $\{G^{adj}_{ij1} \cong G^{adj}_{ij2} \cong...\cong G^{adj}_{ijq}\}_n \subseteq \Gamma^G{}_n.$

A binary position ΩR_n, as *isomorphism class* Γ_n can be interpretable also as an "isomorphism clique", where all the element pairs are mutually isomorphic.

P2.6.2.   Isomorphism class of adjacent graphs $\Gamma^G{}_n$ *is replaceable* with corresponding *isomorphism class of binary graphs* $\{g_{ij1} \cong g_{ij2} \cong...\cong g_{ijq}\}_n \subseteq \Gamma^g{}_n$, that as well characterize the vertex pairs. Identification the elements of position ΩR_n take place by *binary signs* $\pm d.n.m._{ij}$ (or $\pm d.n.m.x._{ij}$) as the identifiers of isomorphism class Γ_n.

P2.6.3.   In the model of structure **SM** is each vertex position ΩV_k related directly with binary positions ΩR_n of its incident edges.

Conclusions 2.1. Comparison the *group theoretic way of orbits recognition* and *semiotic modeling of positions*:

1) The **vertex-** ΩV_k and **binary positions** ΩR_n are recognized in structure model **SM**.
2) The orbits, recognized by group theoretic orbits, and positions, recognized by semiotic modelling, **coincide!**
3) Graphs with different structures can be have one and same group **AutG**, but have different semiotic models **SM**.
4) In case group theoretic treatment the number of permutations of completely symmetric graphs can be increase up to factorial. In case semiotic modelling of this does not happen.
5) In case group theoretic treatment the recognitions of vertex and edge orbits takes place separately and the "non-edge orbits" does not exist. In case semiotic modelling the recognitions of vertex-, pair(+)- and pair(-)orbits take place completely, where structure model **SM** express these in a complex.
6) Up to present considered, that orbit recognition belongs to periphery of graph theory. On the semiotic aspect it is a central problem.

Corollary 2.1. *A position (equivalence class) and orbit is the same.*

Symmetry properties desirable to be distinguish. Symmetry properties depend from the positions.

Definitions 2.1. The *symmetry kinds* of graph structure:

D2.1.1.   Graph with only *one* vertex position ΩV_k we call *vertex symmetric* that also *transitive* called.

For vertex symmetric or transitive graphs:

D2.1.2.   Vertex symmetric graph with only *one* binary position ΩR^+_n is *completely symmetric* or *complete graph.*

Empty graph with only one "non-edge" or pair position (orbit) is also *completely symmetric*.

D2.1.3.   Vertex symmetric graph with only *one* edge position (i.e. binary(+)position) ΩR_n^+ and only *one* "non-edge" position (i.e. binary(−)position) ΩR_n^- we call *bisymmetric graph.*

For example, bisymmetric is Petersen graph (example 2.1).

D2.1.4.   Vertex symmetric graph with *one* edge position (binary(+)position) ΩR_n^+ and *any* "non-edge" positions (binary(−)positions) ΩR_n^- we call *edge symmetric* or *(+)symmetric graph*.

For example, edge symmetric are here Dodecahedra (example 2.2) and Heawood's (example 2.4) graphs. Complement of an *edge symmetric* graph is a *"non-edge"-* or *(−)symmetric graph*. Jointly we call these *mono symmetric graphs.*

D2.1.5.   Vertex symmetric graph with *any* edge positions (binary(+)positions) ΩR_n^+ and *any* "non-edge" positions (pair(−)orbits) ΩR_n^- we call *poly-symmetric graph.*

For example, poly-symmetric are here the graphs on examples 3.2 and 3.8. Transitive graphs exist rarely. Among 156 of 6-elements structures exists such only 8.

For non vertex symmetric graphs:

D2.1.6.   Graph with *more than one* vertex position ΩV_k, whereby at least to one ΩV_k belong at least two elements we call *partially symmetric graph.*

For example, partially symmetric graph showed on examples 1.2 − 1.5. Partial symmetry is a broad form of transition at symmetry to **a**symmetry. Among 156 of 6-elements structures are 140 partially symmetric.

D2.1.7.   Graph where the number of vertices $|V|$ and vertex positions ΩV_k K is equal is a *0-symmetric* or *(completely) asymmetric graph.*

For example, a *0*-symmetric graph showed on example 1.8. In case of little graphs is also an exceptional phenomenon. For example, among 156 of 6-elements structures are *0*-symmetric only 8.

For presentation the symmetry of structure is suitable to use corresponding signs.

Definitions 2.2. *Symmetry signs* of the structure*:*

D2.2.1 A vector with elements $|\Omega V|^m$, where $|\Omega V|$ is the power of a vertex position and m is the number of positions with such power, called *sign of vertex symmetry SVV.*

D2.2.2. A vector with elements $|\Omega R|^m$, where $|\Omega R|$ is the power of a pair position and m is the number of positions with such power, called *sign of pair symmetry SRV.*

Symmetry signs of the graphs on Examples 1.2 – 1.5 coincide, these are $SVV=1^1\,2^1\,3^1$, $SRV=1^1\,2^1\,3^2\,6^1$. The *edge symmetry* is here different, on Example 1.2 it is $SEV=1^1\,3^1\,6^1$.

Symmetry signs give a good possibility to *measuring* of the structure.. To foundation of *symmetry size* is the classical Shannon's formula of *information capacity*. Information capacity is practically a measure of *asymmetry or inner diversity*.

Propositions 2.7. *Measurement* of the inner diversity:

P2.7.1. *Vertex information capacity HV* depends from the number of vertices $|V|$ and the power of vertex positions $|\Omega V_k|$:
$$HV = -_{k=1}\Sigma^K PV_k \log PV_k,$$
where $0 \le PV_k = |\Omega V_k| : |V| \le 1$.

There $minHV = 0 \le HV \le \log|V| = maxHV$, where, if $K=1$, then $HV=0$ and if $K=|V|$, then $HV=\log|V|$.

P2.7.2. *Binary information capacity HR* depends from the number of vertex pairs $|R|$ and the power of binary positions $|\Omega R_n|$:
$$HR = -_{n=1}\Sigma^N PF_n \log PF_n,$$
where $0 \le PF_n = |\Omega R_n| : |R| \le 1$ ja $|R| = [|V|(|V|-1)]:2$.

There $minHR = 0 \le HR \le \log|R| = maxHR$, where, if $N=1$, then $HR=0$ and if $N=|R|$, then $HR=\log|R|$. Edge info capacity HR^+ calculates by the number of edges $|E|$ and the power of edge positions $|\Omega R_n^+|$. "Non-edge" info capacity HR^- calculate by the number of "non-edges" $|R^-|$ and the power of corresponding pair positions $|\Omega R_n^-|$.

Information comes into being on the ground of certain *diversity, i.e. inner distinctness*. Information capacity depends from *quantity of variances*. There where variances no exist, arises a certain "domain of equability", what on the structural aspect a *symmetry class* or *position* mean. Then more exist "domains of equability" or positions, then larger is information capacity **HR** and then smaller is the *symmetry size (value)*.

Proposition 2.8. On the ground of information capacities *HV* and *HR* can be recognize the **symmetry values SV** and **SR** correspondingly:

$$SR = 1 - (HR : \log|R|), \text{ where } 0 \le SR \le 1.$$

The symmetry *value is 1*, if there exist *only one position*; the *value is 0*, if the *number of positions equal to the number of elements*. This give rise to *compare, order and grouping* the graphs with different size by symmetry values By analogy with the value of *binary symmetry* **SR** can be express the *binary(+)symmetry ("edge-symmetry")* **SE**. Symmetry value **SR** is officially called as **regularity.**

Example 2.5. Symmetry-vectors and the symmetry-values of the graphs showed on various examples. Ordered by lessen of the edge symmetry **SE**:

Exm	Sym	K	N	SVV	SV	SRV	HR	SR	SEV	SE	aut	3003PS
2.7.	Bis	1	2	6^1	1.000	$3^1 12^1$	0.722	0.815	3^1	1.000	48	99
2.8.	Bis	1	2	6^1	1.000	$6^1 9^1$	0.971	0.751	6^1	1.000	72	6

1.5.	Prt	3	5	$1^1 2^1 3^1$	0.478	$1^1 2^1 3^2 6^1$	2.106	0.461	3^1	1.000	12	396
1.4.	Prt	3	5	$1^1 2^1 3^1$	0.478	$1^1 2^1 3^2 6^1$	2.106	0.461	$1^1 6^1$	0.789	12	28
1.2.	Prt	3	5	$1^1 2^1 3^1$	0.478	$1^1 2^1 3^2 6^1$	2.106	0.461	$1^1 3^1 6^1$	0.610	12	60
1.3.	Prt	3	5	$1^1 2^1 3^1$	0.478	$1^1 2^1 3^2 6^1$	2.106	0.461	$2^1 3^1$	0.582	12	60
4.2.	Prt	4	9	$1^2 2^2$	0.266	$1^3 2^6$	3.107	0.205	$1^2 2^4$	0.241	2	360
GS76	0-s	6	15	1^6	0	1^{15}	3.907	0	1^8	0	1	336

*

Each position is "naturalizable" in the form of a *position structure* [31, 35, 36].

Definition 2.3. *Position structure GS_n* is a structure that consists of element pairs, which belong to a certain *binary position ΩR_n*. The number of position structure equal to the number of binary positions.

The position structures opens some various "hidden sides" of the structure, that sometimes also "mystical" seems. In principle, the position structures are inevitable, so as the cowering, cliques and others structural attributes, where their identification to a very practical and necessary deemed.

Example 2.6. *Bipartite and semi-symmetric* Folkman's graph *Fol*, its binary signs, structure model and list of its position structures *GS_n*:

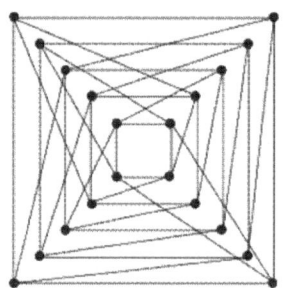

$$A:-4.14.21;\ B:-3.8.10;\ C:-2.6.8;\ D:-2.4.4;\ E:-2.3.2;\ \mathbf{F:+3.6.8.}$$

1 1 1 1 1 1 1 1 1 1	2 2 2 2 2 2 2 2 2 2		u_i	k	s_i
11 12 13 14 15 16 17 18 19 20	1 2 3 4 5 6 7 8 9 10	i	ABCDEF		12
0 -E -E -E -E -E -E -E -E -C	F -B -B F -B F -B -B -B -F	11	061084	1	04
0 -E -E -E -E -E -E -E -C -E	-B F -B -B F -B -B -B F F	12	061084	1	04
0 -E -E -E -E -C -E -E	F -B F -B -B -B -B F F B	13	061084	1	04
0 -E -E -C -E -E -E	-B F -B F -B -B F F -B -B	14	061084	1	04
0 -C -E -E -E -E	-B -B F -B F F F -B -B -B	15	061084	1	04
0 -E -E -E -E	-B -B F -B F F F -B -B -B	16	061084	1	04
0 -E -E -E	-B F -B F -B -B F F -B -B	17	061084	1	04
0 -E -E	F -B F -B -B -B -B F F -B	18	061084	1	04
0 -E	-B F -B -B F -B -B -B F F	19	061084	1	04
0	F -B -B F -B F -B -B -B F	20	061084	1	04
	0 -A -D -D -A -D -A -D -D -D	1	360604	2	40
	0 -A -D -D -A -D -D -D -D	2	360604	2	40
	0 -A -D -D -D -D -D -A	3	360604	2	40
	0 -A -D -D -D -A -D	4	360604	2	40
	0 -D -D -A -D -D	5	360604	2	40
	0 -D -A -A -D	6	360604	2	40
	0 -D -A -A	7	360604	2	40
	0 -D -A	8	360604	2	40
	0 -D	9	360604	2	40
	0	10	360604	2	40

Explanations:
  a) Graph *Fol* decompose correspondingly it's binary positions –*A, –B, –C, –D, –E and **F*** to six *position-structures*:
  b) To binary position –*A* corresponds *position structure* $Fol_{n: -A}$ is **Petersen's graph(!)**. This fact is showed also in partial model $SM_{2.2}$, if there the sign –*A* replaced with Petersen sign +*4.10.15* and –*D* replaced with sign –*2.3.2* then it is equivalent with structure model of Petersen graph (see example 1.2).
  c) To binary position –*B* corresponds *position structure* $Fol_{n=-B}$ turns out to *another semi-symmetric graph*, designed by V. Titov [41] that has also a position structure in the form of *Petersen graph*.
  d) To binary position –*C* corresponds *position structure* $Fol_{n=-C}$ is a graph with ten *components of 2-cliques*.
  e) To binary position –*D* corresponds *position structure* $Fol_{n=-D}$ is the *complement of Petersen graph* (!).
  f) To binary position –*E* corresponds *position structure* $Fol_{n=-E}$ is the *complement of position structure* $Fol_{n=-C}$, i.e. *2-quinta clique*.
  g) To binary position +*F* corresponds *position structure* $Fol_{n=+F}$ is naturally *Folkman graph* self.

The importance of position structures lies in the explaining structural properties, where these also recognize the identical particles of various structures. For example, could be argued that the semi-symmetrical graphs with 20 elements represent a kind of "genetic group" that contains position structures in the form of Petersen graphs. Such relationships between the position structures appear in various ways.

Position structures GS_n opens the different "hidden" sides and particles of its initial structure *GS* If the structure is divided to certain parts, or contain components, cliques, girths, etc., then appear the corresponding attributes in position structures in another forms.

Propositions 2.9. Properties of position structures:
P2.9.1. Position structure is *element symmetric*, i.e. its elements belong to the same position $\Omega V_{k=1}$.
P2.9.2. To the binary(+)position ΩR_n^+ corresponds a position(+)structure GS_n^+ is a *partial structure of GS*; to the binary(–)position ΩR_n^- corresponds a position(–)structure GS_n^- is a *partial structure of complement* $]GS$.
P2.9.3. To each binary(+)position ΩR_n^+ of structure *GS* corresponds the binary(–)position of complement $]GS$ where their *position structures coincides*, $GS_n^+ \equiv]GS_n^-$.
P2.9.4. Some position structure GS_n can be *appear isomorphic with initial structure, GS*, $GS_n \cong GS$ (for example, a position structure of the cube is also cube).
P2.9.5. *Different position structures GS_n of initial structure GS or position structures of different structures* can be *isomorphic* or *coincides*.

By help position structures can be find the same attributes of various structures. For example, Hypercube and Möbius-Kantor graph have some common position structures etc. Under the looking are also the position structures of position structures, i.e. second and high degree position structures.

Propositions 2.10. Properties of high degree position structures:
P2.10.1. A second or high degree position structure can be *isomorphic or coincides with a lower degree position structure or initial structure. Coincidence* of a position structure and initial structure constitutes a *reconstruction* of initial structure.
P2.10.2. High degree position structures *no open more complementary "hidden sides"*, these begin to repeat.
P2.10.3. Formation of high degree position structure is a *convergent process*, it finished with a crop up or reconstruction a low degree or initial structure.

2.3. Relationships between regularity and symmetry properties

Interest for the relationship of regularity and symmetry is not seen, as in case of practical tasks these properties are usually not visible. However, it is with real legitimacy. An interesting relationship is between the bisymmetry and strong regularity, which seems to be "hidden" behind.

A graph said **strongly regular** with parameters (k,a,b) if it is a k-regular incomplete and connected graph such that any two adjacent vertices have exactly $a \geq 0$ common neighbors and any two non-adjacent vertices have $b \geq 1$ common neighbors.

Existence in connected bisymmetric structure exactly two different binary signs, $-d.n_1.q$ and $+d.n_2.q$, mean that by $\pm d=2$ *has each nonadjacent vertex pair exactly n_1-2 common neighbors* and *each adjacent vertex pair n_2-2 common neighbors*. In case $+d>2$ no exist common neighbors. The numbers n-2 of common neighbors can be stay constant also by existence more that two binary signs, i.e. by mono-, poly- and partial symmetries. Thus, strongly regular graphs can be also *mono-, poly- and partial symmetric*.

Proposition 2.11. All the *connected bisymmetric* structures are **strongly regular** as well **girth-** or **clique regular**.

Proposition 2.12. The **connected complement** of a strongly regular structure is also *strongly regular*.

Proposition 2.13. The **complement** of a graph with m equal *disconnected partial cliques* is a **bisymmetric m-partite complete graph**, i.e. it is a **n-m-clique** – and contrariwise.

Size	Graph	Its complement
m	Number of disconnected partial cliques	Number of parts
n	Power of disconnected partial cliques	Power of parts

This mean that the complement of the structure with two disconnected partial cliques is a *bi-clique*, with three disconnected partial cliques is a *tri-clique*, etc.

Example 2.7. Graph **B6-3**, its complement **B6-12**, their binary signs, structure models and measures:

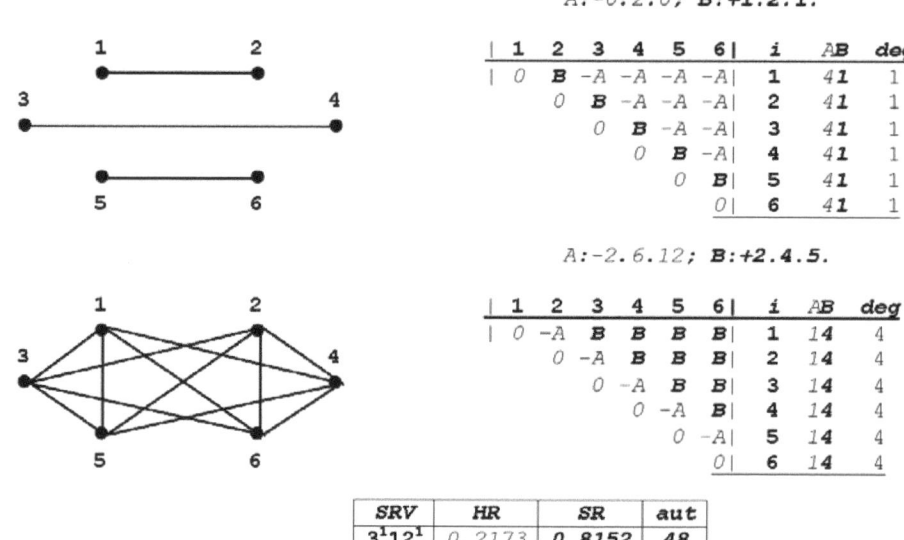

```
                              A:-0.2.0;  B:+1.2.1.

   1              2      | 1   2   3   4   5  6|   i    AB   deg
                         | 0   B  -A  -A  -A -A|   1    41    1
 3              4        |     0   B  -A  -A -A|   2    41    1
                         |         0   B  -A -A|   3    41    1
                         |             0   B -A|   4    41    1
                         |                 0  B|   5    41    1
   5            6        |                    0|   6    41    1


                              A:-2.6.12;  B:+2.4.5.

   1              2      | 1   2   3   4   5  6|   i    AB   deg
                         | 0  -A   B   B   B  B|   1    14    4
 3              4        |     0  -A   B   B  B|   2    14    4
                         |         0  -A   B  B|   3    14    4
                         |             0  -A  B|   4    14    4
                         |                 0 -A|   5    14    4
   5            6        |                    0|   6    14    4
```

SRV	HR	SR	aut
$3^1 12^1$	0.2173	0.8152	48

Explanations: **a)** Graph **B6-3** and its complement **B6-12** are *bisymmetric.* **b)** Graph **B6-3** consist of *three disconnected partial 2-cliques*, it is *2-clique regular.* **c)** Complement **B6-12** is *three partite*, where its parts correspond to 2-cliques of **B6-3**. It is a so called *partite clique*, exactly with a *2-tri-clique*, generally called *n-m-clique*. It is simply sight that all the vertices belong to *triangles*.

Example 2.8. Graph **B6-6,** its complement **B6-9**, their binary signs, structure models and measures:

$$A:-0.2.0; \quad B:+2.3.3.$$

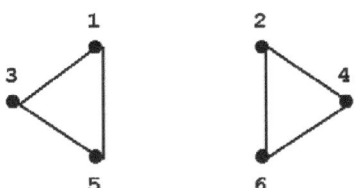

	1	2	3	4	5	6		i	ABC	deg
	0	-A	B	-A	B	-A		1	32	2
		0	-A	B	-A	B		2	32	2
			0	-A	B	-A		3	32	2
				0	-A	B		4	32	2
					0	-A		5	32	2
						0		6	32	2

$$A: -2.5.6; \quad B:+3.6.9.$$

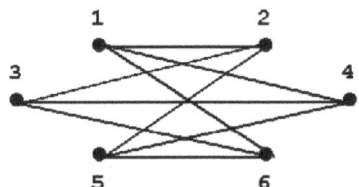

	1	2	3	4	5	6		i	AB	deg
	0	B	-A	B	-A	B		1	23	3
		0	B	-A	B	-A		2	23	3
			0	B	-A	B		3	23	3
				0	B	-A		4	23	3
					0	B		5	23	3
						0		6	23	3

SRV	HR	SR	aut
$6^1 9^1$	0.2923	0.7515	72

Explanations: **a)** Graph **B6-6** and its complement **B6-9** are *bisymmetric.* **b)** Graph **B6-6** consist of *two disconnected partial 3-cliques*, it is *3-clique regular.* **c)** Complement **B6-9** is *bipartite*, where its parts correspond to 3-cliques of **B6-6**. **d)** **B6-9** is a *3-bi-clique* that is *4-girth regular.* **e)** Binary sign *+3.6.9* cover all the *n*=6 vertices and all the *q*=9 edges, it is the *complete invariant* of **B6-9**.

Propositions 2.14. Relationships between *n-m-cliques* and *bisymmetry.*
P2.14.1. All the *n-m-cliques* with equal power *n* and their complements are *bisymmetric.*
P2.14.2. A *n-m-clique* contain an (usual)clique with the power *m*, it is *m-clique regular.*
For example, bi-clique is *2-clique regular,* tri-clique is *3-clique regular,* etc.
P2.14.3. A bisymmetric *n-m-clique* contain $s=n^m$ *usual cliques* with power *m.*
P2.14.4. The number of edges |*E*| of bisymmetric *n-m-clique* equal to the product of quadrate of power n^2 of parts and the number *m* of edges in the usual clique:

n-m-clique			
Symmetry	Power of cliques	Number of cliques	Number of edges
Bisymmetry	*M*	n^m	$E=n^2 m(m-1):2$

Proposition 2.15. Correspondingly to the number of parts we call a *n-m-clique* to *bi-, tri-, quadro-, quinta-, sexta-, septa-, octa-, nona-, deca-,* or *undeca-* etc *-clique.*
Among the graphs with to 20 vertices there exist exactly following *n-m-cliques*:
a) One *n-m-clique* with 4-vertces – *2-biclique* as a complement of disconnected partial 2-cliques;
b) Two *n-m-cliques* with 6 vertices – *2-tri-clique* and *3-bi-clique* as complements of disconnected partial 2- and 3-cliques correspondingly;

c) Two *n-m-cliques* with 8 vertices – **2-quadro-clique** and **4-bi-clique** as complements of disconnected partial 2- and 4-cliques correspondingly;

d) One *n-m-clique* with 9 vertices – **3-tri-clique** as a complement of disconnected partial 3-cliques correspondingly;

e) Two *n-m-cliques* with 10 vertices – **2-quinta-** and **5-bi-clique** as complements of disconnected partial 2- and 5-cliques correspondingly;

f) Four *n-m-cliques* with 12 vertices – **2-sexta-, 3-quadro-, 4-tri-** and **6-bi-clique** as complements of disconnected partial 2-, 3-, 4- and 6-cliques correspondingly;

h) Two *n-m-cliques* with 14 vertices – **2-septa-** and **7-bi-clique** as complements of disconnected partial 2- and 7-cliques correspondingly;

i) Two *n-m-cliques* with 15 vertices – **3-quinta-** and **5-tri-clique** as complements of disconnected partial 3- and 5-cliques correspondingly;

j) Three *n-m-cliques* with 16 vertices – **2-octa-, 4-quadro-** and **8-bi-clique** as complements of disconnected partial 2-, 4- and 8-cliques correspondingly;

k) Four *n-m-cliques* with 18 vertices – **2-nona-, 3-sexta-, 6-tri-** and **9-bi-clique** as complements of disconnected partial 2-, 3-, 6- and 9-cliques correspondingly;

l) Four *n-m-cliques* with 20 vertices – **2-deca-, 4-quinta-, 5-quadro-** and **10-bi-clique** as complements of disconnected partial 2-, 4-, 5- and 10-cliques correspondingly;

In all **27 *n-m-cliques***.

Conclusion 2.2. All the ***n-m-cliques*** are *strongly regular*, but no on the contrary.

In addition of simply constructable ***n-m-cliques*** are recognized following bisymmetric-strongly regular structures: **1)** self-complemented 5-girth; **2)** self-complemented B9-18; **3)** Petersen graph ***PET*** (B10-15); **4)** and its complement ***PETC*** (B10-30); **5)** self-complemented B13-39; **6)** Weisfeiler's B15-45; **7)** and its complement B15-60; **8)** Greenwood's (Clebish's) B16-40; **9)** and its complement B16-80; **10)** Shrikhande B16-48; **11)** and its complement B16-72; **12)** self-complemented B17-68.

Consequently, among the structures with 1 to 20 vertices there exists 27+12=39 bisymmetric + strongly regular + clique- or girth regular graphs

Example 2.9. All the *bisymmetric + strongly regular + clique- or girth regular* graphs with to 20 vertices:

Nr	Notation	deg	SRV	SR	Cmp/prt		Regu-larity	Numbers	Commentary	Pair signs	
					m	n				Binary(-)sign	Binary(+)sign
1	B4-4	2	$2^1 4^1$	0.6448	2p	2	4-girth	-	2-bi-clique	*-2.4.4*	*+3.4.4*
2	B5-5	2	5^2	0.6990	1c	5	5-girth	-	Selfcomplem.	-2.3.2	*+4.5.5*
3	B6-12	4	$3^1 12^1$	0.8152	3p	2	3-clique	8	2-tri-clique	*-2.6.12*	+2.4.5
4	B6-9	3	$6^1 9^1$	0.7515	2p	3	4-girth	-	3-bi-clique	-2.5.6	*+3.6.9*
5	B8-24	6	$4^1 24^1$	0.8769	4p	2	4-clique	16	2-quadro-clique	*-2.8.24*	+2.6.13
6	B8-16	4	$12^1 16^1$	0.7906	2p	4	4-girth	-	4-bi-clique	-2.6.8	*+3.8.16*
7	B9-27	6	$9^1 27^1$	0.8431	3p	3	3-clique	27	3-tri-clique	*-2.8.21*	+2.5.7
8	B9-18	4	18^2	0.8066	3p	3	3-girth	6	Selfcomplem.	-2.4.4	+2.3.3
9	B10-40	8	$5^1 40^1$	0.9084	5p	2	5-clique	32	2-quinta-clique	*-2.10.40*	+2.8.25
10	B10-15	3	$15^1 30^1$	0.8328	1c	10	5-girth	12	Petersen gr.	-2.3.2	*+4.10.15*
11	B10-30	6			1c	10	4-clique	5	Petersen comp.	-2.6.12	+2.5.8
12	B10-25	5	$20^1 25^1$	0.8196	2p	5	4-girth	-	5-bi-clique	-2.7.10	*+3.10.25*
13	B12-60	10	$6^1 60^1$	0.9273	6p	2	6-clique	64	2-sexta-clique	*-2.12.60*	+2.10.41
14	B12-54	9	$12^1 54^1$	0.8868	4p	3	4-clique	81	3-quadro-clique	*-2.11.45*	+2.8.22
15	B12-48	8	$18^1 48^1$	0.8601	3p	4	3-clique	64	4-tri-clique	*-2.10.32*	+2.6.9
16	B12-36	6	$30^1 36^1$	0.8355	2p	6	4-girth	-	6-bi-clique	-2.8.12	*+3.12.36*
17	B13-39	6	39^2	0.8409	1c	1	3-clique	22	Selfcomplem.	-2.5.7	+2.4.5

		deg	SRV	SR		c		s			
18	**B14-84**	12	$7^1 84^1$	*0.9399*	**7p**	2	*7-clique*	128	***2-septa-clique***	*–2.14.84*	*+2.12.61*
19	**B14-49**	7	$42^1 49^1$	*0.8470*	**2p**	7	*4-girth*	-	***7-bi-clique***	*–2.9.14*	*+3.14.49*
20	**B15-90**	12	$15^1 90^1$	*0.9119*	**5p**	3	*5-clique*	243	***3-quinta-clique***	*–2.14.78*	*+2.11.46*
21	**B15-75**	10	$30^1 75^1$	*0.8711*	**3p**	5	*3-clique*	125	***5-tri-clique***	*–2.12.45*	*+2.7.11*
22	**B15-45**	6	$45^1 60^1$	*0.8533*	**1c**	15	*3-clique*	-	***Weisfeiler***	*–2.5.6*	*+2.3.3*
23	**B15-60**	8			**1c**	15	*5-clique*	-	***Weisfeil. comp.***	*–2.6.12*	*+2.6.12*
24	**B16-112**	14	$8^1 112^1$	*0.9488*	**8p**	2	*8-clique*	256	***2-octa-clique***	*–2.16.112*	*+2.14.85*
25	**B16-96**	12	$24^1 96^1$	*0.8955*	**4p**	4	*4-clique*	256	***4-quadro-clique***	*–2.14.72*	*+2.10.33*
26	**B16-40**	5	$40^1 80^1$	*0.8670*	**4p**	4	*4-girth*	-	***Greenwood***	*–2.4.4*	*+3.10.13*
27	**B16-80**	10			**1c**	16	*5-clique*	16	***Greenw. comp.***	*–2.8.24*	*+2.8.22*
28	**B16-48**	6	$48^1 72^1$	*0.8594*	**1c**	16	*4-clique*	-	***Shrikhande***	*–2.4.4*	*+2.4.6*
29	**B16-72**	9			**1c**	16	*4-clique*	-	***Shrikhan comp.***	*–2.8.18*	*+2.6.11*
30	**B16-64**	8	$56^1 64^1$	*0.8557*	**2p**	8	*4-girth*	-	***8-bi-clique***	*–2.10.10*	*+3.16.64*
31	**B17-68**	8	68^2	*0.8589*	**1c**	17	*3-clique*	-	***Selfcomplem.***	*–2.6.11*	*+2.5.7*
32	**B18-144**	16	$9^1 144^1$	*0.9555*	**9p**	2	*9-clique*	512	***2-nona-clique***	*–2.18.144*	*+2.16.113*
33	**B18-135**	15	$18^1 135^1$	*0.9280*	**6p**	3	*6-clique*	729	***3-sexta-clique***	*–2.17.120*	*+2.14.79*
34	**B18-108**	12	$45^1 108^1$	*0.8796*	**3p**	6	*3-clique*	216	***6-tri-clique***	*–2.14.60*	*+2.8.13*
35	**B18-81**	9	$72^1 81^1$	*0.8626*	**2p**	9	*4-girth*	-	***9-bi-clique***	*–2.11.18*	*+3.18.81*
36	**B20-180**	18	$10^1 180^1$	*0.9607*	**10p**	2	*10-clique*	1036	***2-deca-clique***	*–2.20.180*	*+2.18.45*
37	**B20-160**	16	$30^1 160^1$	*0.9169*	**5p**	4	*5-clique*	1924	***4-quinta-clique***	*–2.18.128*	*+2.14.73*
38	**B20-150**	15	$40^1 150^1$	*0.9019*	**4p**		*4-clique*	625	***5-quadro-clique***	*–2.17.105*	*+2.12.46*
39	**B20-100**	10	$90^1 100^1$	*0.8682*	**2p**	10	*4-girth*	-	***10-bi-clique***	*–2.12.20*	*+3.20.100*

Explanations: **a)** The marking of structure show the numbers of vertices and edges. **b)** *deg* – degree. **c)** *SRV* – symmetry vector (Def. 2.2); **d)** *SR* – symmetry value (Prop. 2.8). **e)** *c* – number of components. **f)** *m* – number of parts. **g)** *n* – power of parts. **h)** *s* – number of cliques.

So it is recognized 39 bisymmetric-strongly regular structures with 4 to 20 vertices, mainly on the ground of disconnected partial cliques induced. The results of J. Petersen (B10-15), A.Titov (B13-39), B. Weisfeiler (B15-45), Greenwood-Gleason-Clebish (B16-40) in the realm of bisymmetry are random coincides, because the first be interested on valence-regularity, other on self-complementary, third on strong regularity, fourth on color-conjecture, others on isomorphism testing etc.

The lists of strongly regular graphs are incomplete. For example, in a special list [26] lacked 31 strongly regular graphs with to 20 vertices. A "most complete" list [27] where be given 33 structures, among these also *n-m-cliques* fail unfortunately **25** (B16-96), **29** (B16-72), **33** (B18-135), **34** (B18-108), **37** (B20-160) and **38** (B20-150).

In the "most complete" list of strongly regular graphs are showed all the to 20 vertices *bi-cliques*, as *complete bipartite graphs*, whereby bi-clique with 4 vertices called *square* and with 6 vertices called *unity*. There are also showed all the *2-m-cliques*, that have title *r-cocktail party graphs*, whereby with 6 vertices called *octahedral graph* and with 8 vertices *16-cell graph*. Other *n-m-cliques* called mostly *circular graphs*. There lack five *n-m-cliques* and the complement of a known strongly regular graph

It is touch with partial coincide the bisymmetry and strong regularity. Bisymmetry cover also disconnected structures and strong regularity can be exists also by mono-, poly- and partial symmetry. But the lasts no exist among the structures with to 20 vertices. Semiotic approach was fill the "white blotch" of lists the strongly regular graphs, was pick out the essence of so far ignored clique regularity, and this that the complement of strongly regular graph is also strongly regular.

In such lists exists also many large graphs. For example, on a list [26] to find a graph with 999 vertices:

16	(999, 448, 172, 224)	-	-

<u>Example 2.10.</u> We can in a simple way to induce some bisymmetric, clique- and strongly regular graphs with 999 vertices. In the lists of strongly regular graphs cannot these to find:

Nr	Notation	deg	|E|	SR	Regularity	Commentary	(+)signs
1	**B999-2**	2	999		*3-clique*	333 disconnected partial 3-cliques	*+2.3.3*
2	**B999-996**	996	497502	*0.9989*	*333-clique*	333 3-elementic parts *3-tricent-triginta-tri-clique*	**?**
3	**B999-8**	8	3996		*9-clique*	111 disconnected partial 9-cliques	*+2.9.36*
4	**B999-990**	990	494505	*0.9979*	*111-clique*	111 9-elementic parts *9-cent-undeca-clique*	**?**
5	**B999-110**	110	54945		*111-clique*	9 disconnected partial 111-cliques	*+2.111.6105*
6	**B999-888**	888	443556	*0.9736*	*9-clique*	9 111-elementic parts *111-nona-clique*	**?**
7	**B999-332**	332	165832		*333-clique*	3 disconnected partial 333-cliques	*+2.333.55278*
8	**B999-666**	666	332667	*0.9515*	*3-clique*	3 333-elementic parts *333-tri-clique*	**?**

Explanations: **a)** Strongly regular are there only *n-m*-cliques. **b)** The names of *n-m*-cliques can be for any no please, but others I cannot find.

With very important symmetry properties is graph *Gre* (B16-40) was constructed by Greenwood-Gleason as in any 3-colouring of the edges of the K_{16} without monochromatic triangles, the set of edges of each colour from this graph. It called also Clebish graph.

<u>Example 2.11.</u> Bisymmetric strongly regular Greenwood-Gleason-Clebish graph *Gre* and the structure models of *Gre* and its complement *GreC*:

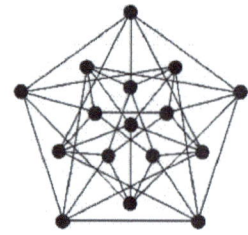

```
A:-2.4.4; B:+3.10.13.
```

```
  1  2  3  4  5  6  7  8  9 10 11 12 13 14 15 16|   i    AB   deg
  0  B -A -A  B -A -A  B -A -A -A  B -A -A  B -A|   1   105    5
     0  B -A -A -A  B -A -A  B -A -A -A  B -A -A|   2   105    5
        0  B -A  B -A -A  B -A -A  B -A -A -A -A|   3   105    5
           0  B -A -A  B -A -A  B -A -A  B -A -A|   4   105    5
              0  B -A -A -A  B -A -A  B -A -A -A|   5   105    5
                 0  B -A -A -A -A -A -A -A  B  B|   6   105    5
                    0  B -A -A  B -A  B -A -A -A|   7   105    5
                       0  B -A -A -A -A -A -A  B|   8   105    5
                          0  B -A -A  B -A  B -A|   9   105    5
                             0  B -A -A -A -A  B|  10   105    5
                                0  B -A -A  B -A|  11   105    5
                                   0  B -A -A  B|  12   105    5
                                      0  B -A -A|  13   105    5
                                         0  B  B|  14   105    5
                                            0 -A|  15   105    5
                                               0|  16   105    5
```

```
                          A:-2.8.24; B:+2.8.22.

   1   2   3   4   5   6   7   8   9  10  11  12  13  14  15  16|   i    A B   deg
   0  -A   B   B  -A   B   B  -A   B   B   B  -A   B   B  -A   B|   1    510    10
       0  -A   B   B   B  -A   B   B  -A   B   B   B  -A   B   B|   2    510    10
           0  -A   B  -A   B   B  -A   B   B  -A   B   B   B   B|   3    510    10
               0  -A   B   B  -A   B   B  -A   B   B  -A   B   B|   4    510    10
                   0  -A   B   B   B  -A   B   B  -A   B   B   B|   5    510    10
                       0  -A   B   B   B   B   B   B   B  -A  -A|   6    510    10
                           0  -A   B   B  -A   B  -A   B   B   B|   7    510    10
                               0  -A   B   B   B   B   B   B  -A|   8    510    10
                                   0  -A   B   B  -A   B  -A   B|   9    510    10
                                       0  -A   B   B   B   B  -A|  10    510    10
                                           0  -A   B   B  -A   B|  11    510    10
                                               0  -A   B   B  -A|  12    510    10
                                                   0  -A   B   B|  13    510    10
                                                       0  -A  -A|  14    510    10
                                                           0   B|  15    510    10
                                                               0|  16    510    10
```

Common invariants and measures of graph and its complement:

Symmetry	$|V|$	$|R|$	K	N	SVV	SV	SRV	HR	SR
Bisymmetry	16	120	1	2	16^1	1.000	$40^1 80^1$	0.2762	0.8670

Distinguishing invariants and measures:

G	$|E|$	k	N^+	N^-	P	CL	MC	DM	SEV^+	SE^+	TRA	BRA
B16-40	40	1	1	1	2	2	4	2	40^1	1.000	0	0
B16-80	80	1	1	1	2	5	3	2	80^1	1.000	1.000	0

Explanations: **a)** The **bisymmetric** and **strongly regular** structure **Gre** is correspondingly to binary(+)sign **+3.10.13** (*a complete invariant!*) **4-girth regular**, that mean *partiting*. This appear also to **4-partite** with incompletely connected parts on *4-elemental bases*. **b)** It is no quadroclique. **c)** The parts are *variety*, where, for example one variant is **A=5,8,12,15; B=3,7,10,14; C=1,4,9,16;** and **D=2,6,11,13**:

	A	**B**	**C**	**D**
A	0	4	6	10
B		0	10	6
C			0	4
D				0

d) From 4-elemintic parts of **Gre** conclude the **4-clique regularity** of variety cliques of complement **GreC**. **e)** On the other hand, in case of each vertex of **Gre** its 5 adjacent vertices no have between themselves adjacencies (edges), from which conclude also a **5-clique-regularity** of complement **GreC**. We can in **GreC** to fix 16 different 5-cliques, such as (beginning at the adjacent vertices of first vertex of **Gre**) **2,5,8,12,15; 1,3,7,10,14;** ... to ending with **6,8,10,12,14**.

Conclusion 2.3. Semiotic approach discovers some new strongly- and clique regular structures.

3. STRUCTURE MODEL AND ISOMORPHISM PROBLEM

Structure model is also a ***canonical presentation*** of a graph. The problem of canonical presentation was established probably by Lazlo Babai [1, 2] in 1977th. It means the presentation of graph in a certain form, preferably *with exactness of isomorphism.* ***Isomorphism problem*** consists in design an algorithm that recognizes the isomorphism of two objects. ***Isomorphism*** (Greek word ισοσ – same; μορφε – form) constitutes *a one-to-one correspondence* between *structures* of objects [20, 25]. Such a one-to-one correspondence can only exist between abstract, idealized objects, which preserve the structure, i.e. relations, ordering, topology etc., of the systems.

3.1. Structure model as canonical presentation of a graph

As a rule, graphs canonized on the basis of *polynomials, spectra* [5], *3-cubecodes* [15] and other *global invariants* [9]. Unfortunately, such canonization does not contain the necessary information about the structure of a graph, this is not modeling. It is suggested to use also *local invariants*, such as *density, paths, cycles, cliques* and other [46]. We show that the binary signs are suitable local invariants and structure model can be considered as the "text" of structure.

<u>Proposition 3.1.</u> Structure model **SM** is a ***canonical presentation*** of the structure with exactness of *binary signs, structural attributes, positions and isomorphism.*

<u>Example 3.1.</u> Structure of Boris Weisfeiler's [45, p. 166 (a)] a *strongly regular* graph ***Wei*** recognizable and presentable on the basis of basic binary signs:

```
A:-2.8.20; B:-2.8.19; C:-2.8.18;
D:+2.7.13; E:+2.7.14; F:+2.7.15.
```

```
|1  1| 2  2|3  3|4  4| 5| 6  6|7  7|8  8| 9  9|10 10|11|12|13|14 14|15|        u_i   k
20 24|12 14|1  2|9 19| 6|10 16|8 18|4  7|11 17|13 15|23| 3|22|21 25| 5|  i ABCDEF   *
|0  F| C  C|C  B|F  C| C| B  F|C  E|F  C| E  F| E  C| B| F| F| F  C| F| 20 039039 1
 0| C  C|B  C|C  F| C| F  B|E  C|C  F| F  E| C  E| B| F| F| C  F| F| 24 039039 1
  | 0  F|F  C|C  C| B| F  C|B  F|B  F| E  C| F  E| F| C| F| F  C| E| 12 039039 2
   0|C  F|C  C|B  C| F| F  B|F  B| C  E| E  F| F| C| F| C  F| E| 14 039039 2
    |0  F| C  E|F  C|F  B| F| C  I| F  C| C  F| B| C| C| E| 1 039039 3
     0|C  F| E|  C| F  B|F  E|F  C| F  C| F| C| F| B| C| E| 2 039039 3
      |0  F| E| F| C| F  E|F  C| B| F  F| C| F| B| E| C| B| C| 9 039039 4
       0| E| C| F  E|F  C| F| F| B| C| F| F| B| E| B| C| C| 19 039039 4
        | 0| C| C  B|B  B| B| E  E| F| F| F| C| F| F| C| 6 066066 5
         | 0| B  F|E  F| E| B  E|B  E| B| B| C| E| E| C| 10 066066 6
          0|E  F|E  F| E| B  E|B  B| B| C| E| E| C| 16 066066 6
           |0  C|F  B| F| B  C| C| E| C| B| C| E| E| 8 066066 7
            0|B  F| B| F| C| C| E| C| B| C| E| E| 18 066066 7
             |0  C| C| C| B| E| C| E| E| E| B| B| 4 066066 8
              0| C| C| E| B| C| E| E| B| E| B| 7 066066 8
               | 0| B| E| F| C| B| B| E| C| F| 11 066066 9
                0| F| E| C| B| B| C| E| B| F| 17 066066 9
                 | 0| B| B| C| E| B| F| B| 13 066066 10
                  0| B| C| E| F| B| B| 15 066066 10
                   | 0| E| D| F| F| E| 23 066147 11
                    | 0| E| F| F| D| 3 066147 12
                     | 0| B| B| B| 22 093174 13
                      | 0| E| A| 21 147066 14
                       0| A| 25 147066 14
                        | 0| 5 255174 15
```

General invariants and measures (values):

Symmetry	|V|	|R|	K	N	SVV	SV	SRV	HR	SR
Partial symmetry	25	300	15	154	$1^5 2^{10}$	0.1723	$1^{20} 2^{128} 4^6$	2.1576	0.1290

Specified invariants and measures of **Wei** and its complement **WeiC**:

G	|E|	k	N⁺	N⁻	P	CL	Girth	DM	SEV⁺	SE	TRA	BRA
WEI	150	1	80	74	6	4	3	2	$1^{12}2^{67}4^{1}$	0.1310	1.000	0
WEIC	150	1	74	80	6	4	3	2	$1^{8}2^{61}4^{5}$	0.1494	1.000	0

Explanations:

a) From structure model read out that **Wei** is ***partially symmetric, strongly regular, triangular, 2-distance- and 12-degree regular.***

b) On the ground of only ***six binary signs*** is the 25×25 structure model decomposed by help ***u***- and ***s***-vectors to ***15 vertex positions (orbits)*** and ***115 partial models*** $SM_{ki,kj}$.

c) 150 "non edges" of **Wei** form ***74 binary(–)positions***, where –*A* form a position with two elements, –*B* form 33 positions, among these 4 with one element and 29 with two elements, –*C* form 40 positions, 4 with one, 31 with two and 5 with four elements.

d) 150 edges of **Wei** form ***80 binary(+)positions***, where +*D* form 2 positions with one element, +*E* 32 positions, among these 4 with one, 27 with two and one with four elements, and +*F* form 46 position, 6 with one and 40 with two elements.

e) B. Weisfeiler [45] is one of these, who find that vertex orbits are essential attributes of graph structure. But the binary orbits he yet not perceives. He had constructed some strongly regular graphs which has grounded on the same pair signs, but are no isomorphic. On structural aspect: these differ from decompositions to binary positions.

f) Graph **Wei** and its complement **WeiC** is ***triangular, 2-distance-*** and ***12-degree regular.*** Complement **WeiC** is also ***strongly regular.***

Now is suitable to present a graph, where the recognition of its binary positions needed adjusted identification (Prop. 1.3).

Example 3.2. A simple ***polysymmetric*** graph **Tev**, its *basic binary signs and structure model*:

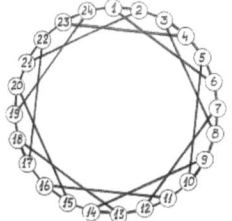

A:–5.18.23; B:–4.9.10; C:–4.8.8; D:–4.7.7; E:–3.8.9; F: –3.3.6; G:–3.4.3; H:–2.3.2;
I:+5.10.12; J:+5.12.15; K:+5.14.18.

```
  1   2   3   4   5   6   7   8   9  10  11  12  13  14  15  16  17  18  19  20  21  22  23  24|   i   ABCDEFGHIJK  k
  0  +K   H   E   H  +I   H   F   B   G   D   G   C   A   D   A   B   G   H   F   H  +J|   1   22121336111   1
      0  +J   H   F   H   F   H   G   B   A   D   A   C   G   D   G   B   F   H  +I   H   E   H|   2   22121336111   1
          0  +K   H   E   H  +I   H   F   B   G   D   G   C   A   D   A   B   G   H   F   H   F|   3   22121336111   1
              0  +J   H   F   H   F   H   G   B   A   D   A   C   G   D   G   B   F   H  +I   H|   4   22121336111   1
                  0  +K   H   E   H  +I   H   F   B   G   D   G   C   A   D   A   B   G   H   F|   5   22121336111   1
                      0  +J   H   F   H   F   H   G   B   A   D   A   C   G   D   G   B   F   H|   6   22121336111   1
                          0  +K   H   E   H  +I   H   F   B   G   D   G   C   A   D   A   B   G|   7   22121336111   1
                              0  +J   H   F   H   F   H   G   B   A   D   A   C   G   D   G   B|   8   22121336111   1
                                  0  +K   H   E   H  +I   H   F   B   G   D   G   C   A   D   A|   9   22121336111   1
                                      0  +J   H   F   H   F   H   G   B   A   D   A   C   G   D|  10   22121336111   1
                                          0  +K   H   E   H  +I   H   F   B   G   D   G   C   A|  11   22121336111   1
                                              0  +J   H   F   H   F   H   G   B   A   D   A   C|  12   22121336111   1
                                                  0  +K   H   E   H  +I   H   F   B   G   D   G|  13   22121336111   1
                                                      0  +J   H   F   H   F   H   G   B   A   D|  14   22121336111   1
                                                          0  +K   H   E   H  +I   H   F   B   G|  15   22121336111   1
                                                              0  +J   H   F   H   F   H   G   B|  16   22121336111   1
                                                                  0  +K   H   E   H  +I   H   F|  17   22121336111   1
                                                                      0  +J   H   F   H   F   H|  18   22121336111   1
                                                                          0  +K   H   E   H  +I|  19   22121336111   1
                                                                              0  +J   H   F   H|  20   22121336111   1
                                                                                  0  +K   H   E|  21   22121336111   1
                                                                                      0  +J   H|  22   22121336111   1
                                                                                          0  +K|  23   22121336111   1
                                                                                              0|  24   22121336111   1
```

Explanation: Graph *Tev* has eleven *sign structures*, among these three sign(+)structures (i.e. substructures of *Tev*) and eight sign(–)structures (i.e. substructures of complement]*Tev*).

We show here two of these.

Example 3.3. **Sign structures** by sign *F: –3.3.6*, **Tev**$_{p= -F}$ and by *A: –5.18.23*, **Tev**$_{p= -A}$ (these not yet the position structures):

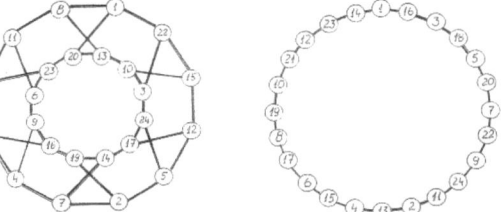

Explanation: Known, that the basic binary signs may not always be complete identifiers of vertex pairs, but on the ground of sign structures can be perfect the basic binary signs up to binary positions.

We show here the using of *productive binary signs* (IIA). The products E^n of small degree no give perfect information about the binary positions. In the case of graph *Tev* are the binary positions recognizable on the degrees $n=6$ and $n=7$ of matrix products E^n.

Example 3.4. *Union of the matrix-products* $\underline{E^{n=6}}$ *and* $E^{n=7}$ *recognize the binary positions of* *Tev*:

i	1	2	3	4	5	6	7	8	9	10	11	12	13	14	15	16	17	18	19	20	21	22	23	24
1	0	258	84	243	75	239	65	191	42	150	33	130	32	107	33	108	42	139	65	173	75	201	84	248
2	258	0	248	84	201	75	173	65	139	42	108	33	107	32	130	33	150	42	191	65	239	75	243	84
3	84	248	0	258	84	243	75	239	65	191	42	150	33	130	32	107	33	108	42	139	65	173	75	201
4	243	84	258	0	248	84	201	75	173	65	139	42	108	33	107	32	130	33	150	42	191	65	239	75
5	75	201	84	248	0	258	84	243	75	239	65	191	42	150	33	130	32	107	33	108	42	139	65	173
6	239	75	243	84	258	0	248	84	201	75	173	65	139	42	108	33	107	32	130	33	150	42	191	65
7	65	173	75	201	84	248	0	258	84	243	75	239	65	191	42	150	33	130	32	107	33	108	42	139
8	191	65	239	75	243	84	258	0	248	84	201	75	173	65	139	42	108	33	107	32	130	33	150	42
9	42	139	65	173	75	201	84	248	0	258	84	243	75	239	65	191	42	150	33	130	32	107	33	108
10	150	42	191	65	239	75	243	84	258	0	248	84	201	75	173	65	139	42	108	33	107	32	130	33
11	33	108	42	139	65	173	75	201	84	248	0	258	84	243	75	239	65	191	42	150	33	130	32	107
12	130	33	150	42	191	65	239	75	243	84	258	0	248	84	201	75	173	65	139	42	108	33	107	32
13	32	107	33	108	42	139	65	173	75	201	84	248	0	258	84	243	75	239	65	191	42	150	33	130
14	107	32	130	33	150	42	191	65	239	75	243	84	258	0	248	84	201	75	173	65	139	42	108	33
15	33	130	32	107	33	108	42	139	65	173	75	201	84	248	0	258	84	243	75	239	65	191	42	150
16	108	33	107	32	130	33	150	42	191	65	239	75	243	84	258	0	248	84	201	75	173	65	139	42
17	42	150	33	130	32	107	33	108	42	139	65	173	75	201	84	248	0	258	84	243	75	239	65	191
18	139	42	108	33	107	32	130	33	150	42	191	65	239	75	243	84	258	0	248	84	201	75	173	65
19	65	191	42	150	33	130	32	107	33	108	42	139	65	173	75	201	84	248	0	258	84	243	75	239
20	173	65	139	42	108	33	107	32	130	33	150	42	191	65	239	75	243	84	258	0	248	84	201	75
21	75	239	65	191	42	150	33	107	32	130	33	108	42	139	65	173	75	201	84	248	0	258	84	243
22	201	75	173	65	139	42	108	33	107	32	130	33	150	42	191	65	239	75	243	84	258	0	248	84
23	84	243	75	239	65	191	42	150	33	130	32	107	33	108	42	139	65	173	75	201	84	248	0	258
24	248	84	201	75	173	65	139	42	108	33	107	32	130	33	150	42	191	65	239	75	243	84	258	0

Explanation: All the identifiers of vertex pairs (i.e. binary positions) here are complete. The matrix products $\underline{E^{n=6}}$ and $E^{n=7}$ are here united, because in both case exist the zero values.

It suitable to associate the basic binary signs with the results of matrix product E^n of this graph:

Example 3.5. Associating the basic and productive binary signs:

	1		2	3	4	5	6			7			8			9	10	11
2	-A		-B	-C	-D	-E	-F			-G			-H			+I	+J	+K
3	1	2	3	4	5	6	7	8	9	10	11	12	13	14	15	16	17	18
4	108	107	42	32	33	243	191	201	173	150	139	130	65	75	84	**239**	**248**	**258**
5	-A1	-A2	-B	-C	-D	-E	-F1	-F2	-F3	-G1	-G2	-G3	-H1	-H2	-H3	+I	+J	+K
6	1	1	2	1	2	1	1	1	1	1	1	1	2	2	2	1	1	1

34

Explanations: 1 – the ordering number of basic binary signs; 2 – basic binary signs (see example 2); 3 – the ordering number of productive binary signs; 4 – productive binary signs (see example 3.4); 5 – marking of productive binary signs (see example 3.6); 6 – the last row is there the *frequency vector* for all the rows (vertices) of structure model. The number of basic binary signs is 11, the number of complete binary signs is 18. Perfected binary sign constitutes a quintuplet $\pm d.n.q.e^{n}_{ij}$, where the last represents the perfecting (see the fives row).

Example 3,6. The *complete structure model* **SM*** of *Tev*:

```
 1  2   3   4   5   6   7   8   9  10  11  12  13  14  15  16  17  18  19  20  21  22  23  24|  i   k
 0  K  H3   E  H2   I  H1  F1   B  G1   D  G3   C  A2   D  A1   B  G2  H1  F3  H2  F2  H3   J|  1   1
    0   J  H3  F2  H2  F3  H1  G2   B  A1   D  A2   C  G3   D  G1   B  F1  H1   I  H2   E  H3|  2   1
        0   K  H3   E  H2   I  H1  F1   B  G1   D  G3   C  A2   D  A1   B  G2  H1  F3  H2  F2|  3   1
            0   J  H3  F2  H2  F3  H1  G2   B  A1   D  A2   C  G3   D  G1   B  F1  H1   I  H2|  4   1
                0   K  H3   E  H2   I  H1  F1   B  G1   D  G3   C  A2   D  A1   B  G2  H1  F3|  5   1
                    0   J  H3  F2  H2  F3  H1  G2   B  A1   D  A2   C  G3   D  G1   B  F1  H1|  6   1
                        0   K  H3   E  H2   I  H1  F1   B  G1   D  G3   C  A2   D  A1   B  G2|  7   1
                            0   J  H3  F2  H2  F3  H1  G2   B  A1   D  A2   C  G3   D  G1   B|  8   1
                                0   K  H3   E  H2   I  H1  F1   B  G1   D  G3   C  A2   D  A1|  9   1
                                    0   J  H3  F2  H2  F3  H1  G2   B  A1   D  A2   C  G3   D| 10   1
                                        0   K  H3   E  H2   I  H1  F1   B  G1   D  G3   C  A2| 11   1
                                            0   J  H3  F2  H2  F3  H1  G2   B  A1   D  A2   C| 12   1
                                                0   K  H3   E  H2   I  H1  F1   B  G1   D  G3| 13   1
                                                    0   J  H3  F2  H2  F3  H1  G2   B  A1   D| 14   1
                                                        0   K  H3   E  H2   I  H1  F1   B  G1| 15   1
                                                            0   J  H3  F2  H2  F3  H1  G2   B| 16   1
                                                                0   K  H3   E  H2   I  H1  F1| 17   1
                                                                    0   J  H3  F2  H2  F3  H1| 18   1
                                                                        0   K  H3   E  H2   I| 19   1
                                                                            0   J  H3  F2  H2| 20   1
                                                                                0   K  H3   E| 21   1
                                                                                    0   J  H3| 22   1
                                                                                        0   K| 23   1
                                                                                            0| 24   1
```

Explanations:
a) Graph *Tev* is *3-degree-* and *6-girth-regular*, its *complement TevC* is *20-degree-, 2-distance-, 3-girth-regular* and also *polysummetric*.
b) From 6-girth regularity concludes that *Tev* is **bipartite**, in present case parts with even- and odd-numbered vertices.
c) As *Tev* is bipartite, but not *bi-clique*, then its complement *TevC* consists of two mutually connected *12-cliques* and is thus *12-clique-regular*. These cliques correspond to parts of *Tev*.
d) The number N of *position-* and *adjacent structures* is 18, their powers coincide in cases *Tev* and its complement *TevC*.
e) 23x24:2 = 276 vertex pairs of *Tev* form 18 **binary positions**, where by 240 "non-edges" be formed 15 binary(–)positions with 12 and 24 elements. 36 adjacent vertex pairs form three binary(+)positions, *+I, +J* and *+K*, with 12 elements and are recognized by the basic binary signs.
f) *Position-structures* of *Tev* are mostly **bisymmetric (two binary positions), 2-clique-regular** and mutually *isomorphic*. Position structures by *–B, –D, –H1, –H2* and *–H3* constitute *girths*.
g) 276 possible *adjacent graphs* converged to 15 *adjacent super-structures* and to three *adjacent sub-structures*.

The *basic binary signs* not lose its meaning, these characterize the relationships between vertices, the belonging of vertex pairs to (assemblage of) paths or girths with corresponding size etc. These are needed for characterizing of the structure as a whole.

Conclusion 3.1. *Time complexity of structure's recognition* depends only at the number of vertex pairs.

3.2. Structural equivalence and graph isomorphism

Isomorphism is *an invertible morphism*, which has *an opposite morphism*, such that their product is *the unity morphism*. A topological isomorphism is called a *homeomorphism*.

The *graph isomorphism problem* first came into prominence in 1857, when Arthur Cayley [4] reported his research on organic isomers. On structural aspect is it the problem of comparing the structural models. We demonstrate that the structure model and isomorphism problem are closely related.

Isomorphic graphs have the same structure, which is expressed in the form of structural equivalence of models.

Proposition 3.2. On the relationships between isomorphism and structural equivalence of graphs:
1) Isomorphism is a one-to-one correspondence between elements where an isomorphic mapping from graph G_A to graph G_B is a bijection $\varphi: V_A \rightarrow V_B$ [1].
2) Isomorphism recognition does not recognize the structure, but the structure model recognizes the structure with exactness up to isomorphism [31].
3) Structural equivalence is a coincidence or bijection on the level of binary signs, binary- and element positions [34].
4) Recognition of the positions by the structure model is more effective than detecting the orbits on the ground of the group *AutG*.

Example 3.7. Graphs G_A and G_B, their basic binary signs and structure models SM_A and SM_B:

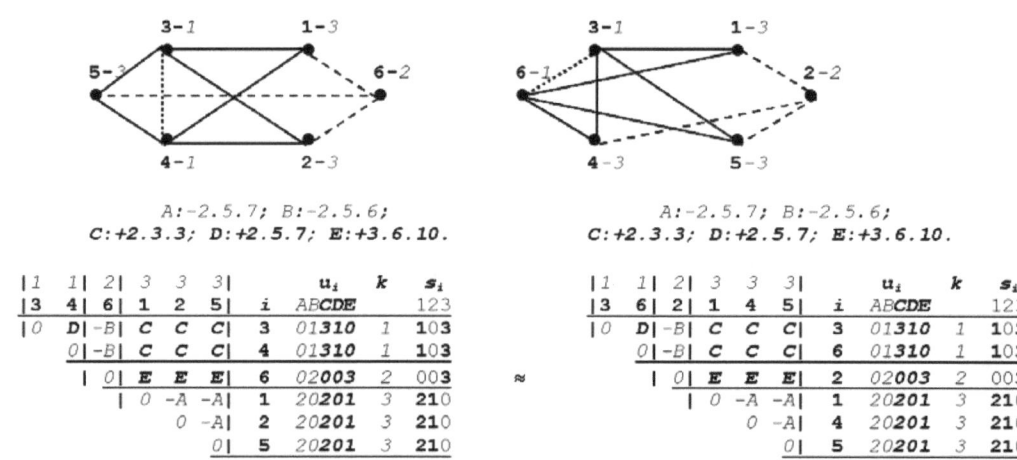

Left model (SM_A):
```
A:-2.5.7;  B:-2.5.6;
C:+2.3.3;  D:+2.5.7;  E:+3.6.10.

|1  1| 2| 3  3  3|        uᵢ     k    sᵢ
|3  4| 6| 1  2  5|   i   ABCDE        123
|0  D|-B| C  C  C|   3   01310   1    103
   0|-B| C  C  C|   4   01310   1    103
     | 0| E  E  E|   6   02003   2    003
     |  0 -A -A|   1   20201   3    210
          0 -A|   2   20201   3    210
            0|   5   20201   3    210
```

$\approx$

Right model (SM_B):
```
A:-2.5.7;  B:-2.5.6;
C:+2.3.3;  D:+2.5.7;  E:+3.6.10.

|1  1| 2| 3  3  3|        uᵢ     k    sᵢ
|3  6| 2| 1  4  5|   i   ABCDE        123
|0  D|-B| C  C  C|   3   01310   1    103
   0|-B| C  C  C|   6   01310   1    103
     | 0| E  E  E|   2   02003   2    003
     |  0 -A -A|   1   20201   3    210
          0 -A|   4   20201   3    210
            0|   5   20201   3    210
```

Explanations:
a) Different graphs G_A and G_B have equivalent structure models $SM_A \approx SM_B$! This means that the structures are *equivalent* and the graphs *isomorphic* $G_A \cong G_B$.
b) The element pairs are divided to *five binary positions* ΩR_n, wherein the adjacent elements or "edges" to *three binary(+)positions (full line, a dotted, dashed-line)* that coincides with binary signs *C, D, E* correspondingly. The structural elements are divided to *three positions* ΩV_k.
c) The column u_i of model consists of the *frequency vectors*, which for the element i show its relations with other elements. On the basis of vectors u_i are arranged the positions in model.
d) The column s_i of model consists of the *position vectors* that represent the connections of element i with elements in corresponding positions k. If on the framework of frequency vectors arises differences of position vectors, then by lasts does a complementary partition into classes.

Proposition 3.3. For recognition the equivalence of structure models *A* and *B* is necessary and sufficient to establish: 1) coincidence of the *sequences of binary signs* $\{\pm d.n.q._{ij}\}_A$ and $\{\pm d.n.q._{ij}\}_B$; 2) coincidence of the *frequency vectors* $\{u_i\}_A$ and $\{u_i\}_B$; 3) coincidence of the *position vectors* $\{s_i\}_A$ and $\{s_i\}_B$.

In following we look the models of the especially for isomorphism testing constructed two *graphs*.

Example 3.8. Poly-symmetric graphs ***Pra***$_A$ and ***Pra***$_B$, their basic and adjusted binary signs and models:

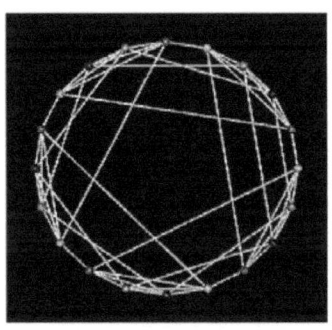 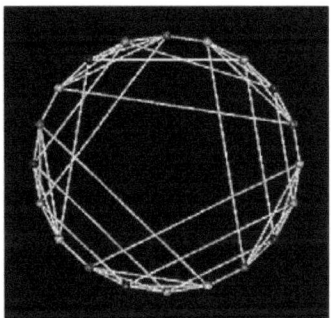

Common basic pair signs of ***Pra***$_A$ and ***Pra***$_B$:

$$A:-3.8.10;\quad B:-3.6.7;\quad C:-2.4.4;\quad D:-2.3.2;\quad \underline{E:+2.4.6};\quad F:+3.8.16.$$

Adjusted by matrix product $E^{n=5}$ binary signs and complete structure model of ***Pra***$_A$:

Marking the basic binary signs	0	-A	-B	-C		-D	E		F
Productive binary signs e^b	180	125	110	165	160	80	231	233	210
Adjusted binary signs	0	-A	-B	-C1	-C2	-D	E1	E2	F

1	2	3	4	5	6	7	8	9	10	11	12	13	14	15	16	17	18	19	20	*i*	*ABCCDEEF*	*k*
0	E2	E1	E1	F	C2	C1	C1	F	C2	C1	C1	D	A	B	B	D	A	B	B	1	24422212	1
	0	E1	E1	C2	F	C1	C1	C2	F	C1	C1	A	D	B	B	A	D	B	B	2	24422212	1
		0	E2	C1	C1	F	C2	C1	C1	F	C2	B	B	D	A	B	B	D	A	3	24422212	1
			0	C1	C1	C2	F	C1	C1	C2	F	B	B	A	D	B	B	A	D	4	24422212	1
				0	E2	E1	E1	D	A	B	B	F	C2	C1	C1	A	D	B	B	5	24422212	1
					0	E1	E1	A	D	B	B	C2	F	C1	C1	D	A	B	B	6	24222212	1
						0	E2	B	B	D	A	C1	C1	F	C2	B	B	A	D	7	24222212	1
							0	B	B	A	D	C1	C1	C2	F	B	B	D	A	8	24222212	1
								0	E2	E1	E1	A	D	B	B	F	C2	C1	C1	9	24222212	1
									0	E1	E1	D	A	B	B	C2	F	C1	C1	10	24222212	1
										0	E2	B	B	A	D	C1	C1	F	C2	11	24222212	1
											0	B	B	D	A	C1	C1	C2	F	12	24222212	1
												0	E2	E1	E1	C2	F	C1	C1	13	24222212	1
													0	E1	E1	F	C2	C1	C1	14	24222212	1
														0	E2	C1	C1	C2	F	15	24222212	1
															0	C1	C1	F	C2	16	24222212	1
																0	E2	E1	E1	17	24222212	1
																	0	E1	E1	18	24222212	1
																		0	E2	19	24222212	1
																			0	20	24222212	

Adjusted by matrix product $E^{n=7}$ binary signs and complete structure model of Pra_B:

Basic binary signs	0	-A	-B		-C			-D	E		F
Productive signs e^7	4410	3437	3276	3277	4081	4088	4011	3010	4831	4803	4445
Adjust. binary signs	0	-A	-B1	-B2	-C1	-C2	-C3	-D	E1	E2	F

```
|1  2  3  4  5  6  7  8  9 10 11 12 13 14 15 16 17 18 19 20|  i   ABBCCCDEEF  k
|0 E1 E2 E1  F C1 C2 C3  F C3 C2 C1  D B2 B1  A  D  A B1 B2|  1   2222222212  1
   0 E1 E2 C3  F C1 C2 C1  F C3 C2  A  D B2 B1 B2  D  A B1|  2   2222222212  1
      0 E1 C2 C3  F C1 C2 C1  F C3 B1  A  D B2 B1 B2  D  A|  3   2222222212  1
         0 C1 C2 C3  F C3 C2 C1  F B2 B1  A  D  A B1 B2  D|  4   2222222212  1
            0 E1 E2 E1  D  A B1 B2  F C1 C2 C3  A  D B2 B1|  5   2222222212  1
               0 E1 E2 B2  D  A B1 C3  F C1 C2 B1  A  D B2|  6   2222222212  1
                  0 E1 B1 B2  D  A C2 C3  F C1 B2 B1  A  D|  7   2222222212  1
                     0  A B1 B2  D C1 C2 C3  F  D B2 B1  A|  8   2222222212  1
                        0 E1 E2 E1  A B1 B2  D  F C3 C2 C1|  9   2222222212  1
                           0 E1 E2  D  A B1 B2 C1  F C3 C2| 10   2222222212  1
                              0 E1 B2  D  A B1 C2 C1  F C3| 11   2222222212  1
                                 0 B1 B2  D  A C3 C2 C1  F| 12   2222222212  1
                                    0 E1 E2 E1 C3  F C1 C2| 13   2222222212  1
                                       0 E1 E2 C2 C3  F C1| 14   2222222212  1
                                          0 E1 C1 C2 C3  F| 15   2222222212  1
                                             0  F C1 C2 C3| 16   2222222212  1
                                                0 E1 E2 E1| 17   2222222212  1
                                                   0 E1 E2| 18   2222222212  1
                                                      0 E1| 19   2222222212  1
                                                         0| 20   2222222212  1
```

Explanations:

a) Pra_A and Pra_B both are **5-degree-, 4-girth-, 4-clique** regular and have *six common basic binary signs*. **4-clique regularity** expressed by existence the five *4*-cliques, what are in structure model showy as signs **E**.

b) Pra_A differ at Pra_B by the number of adjusted binary positions, *eight* and *ten* correspondingly. Consequently, structures Pra_A and Pra_B are **non equivalent** and its graphs **non isomorphic.**

c) Both graphs have three binary(+)positions **E1, E2** and **F** with power 20.

d) Graph Pra_A has five pair(−)positions: by −A, −C2, and −D with power 20, and by −B and −C1 with power 40.

e) The **complement** $PraC_A$ of Pra_A has pair signs −A:-2.14.68, -B:-2.12.47, C:+2.10.35, D:+2.10.36, E:+2.11.44, F:+2.12.48 and is **triangular** and **14-degree regular**.

f) Graph Pra_B has seven pair(−)positions with power 20.

On the graphs Pra_A and Pra_B give interest also their **position structures** that are more than *2-degee-regular*. Such are position structures of Pra_A by positions −B and −C1. For example, position structure $Pra_{A(-B)}$ is (+)**symmetric, 5-partite** and **4-girth regular**, where the parts of $Pra_{A(-B)}$ correspond to *4*-cliques of Pra_A: **I** – vertices **1,2,3,4; II** – vertices **5,6,7,8; III** – **9,10,11,12; IV** – **13,14,15,16; V** – **17,18,19,20.**

Conclusion 3.2. *Time complexity of ascertaining the structural equivalence* depends only at the number of vertex pairs and is polynomial.

3.3. Isomorphism recognition of strongly symmetric graphs

It is possible to construct such *bisymmetric and strongly regular graphs* that have very small binary graphs in case of large number of vertices. We call these ***strongly symmetric graphs.***

Constructed by M. Nechepurenko, M. Klin [19] et al in Siberian *strongly symmetric graphs **Sib_A*** and ***Sib_B*** with 40 vertices have ***common binary signs***: *–A:–2.6.8* (complement has *+B:+2.20.142*) and *+B:+2.4.6* (the complement has *–A:–2.20.144*). From binary signs conclude that *Sib_A* and *Sib_B* are *4-clique-, 2-distance-* and *12-degree regular*. From coincidence the binary signs of *Sib_A* and *Sib_B* conclude the *coincidence of the symmetry properties*.

As in case of strongly regular graphs the involution identification *IIA* no works, we must use another methods of deep-identification. By *high identification* (P1.3.1) the *second degree binary signs of Sib_A* and *Sib_B* are *$–A^{m=2}=–3.18.48$* and *$+B^{m=2}=+3.20.64$*, and anew coincide. A binary graph of third degree *$g_{ij}{}^{m=3}$* no arise, it is empty $\varnothing$. Now must be form by help the *local identification method* (P1.3.2) *local structure models **$SM_{ij}{}^{m=2}$*** for *second degree binary graphs $g_{ij}{}^{m=2}$* of *Sib_A* and *Sib_B*. For this we open in both graphs a binary graph *$g_{ij}{}^{m=2}$*, such that correspond to binary sign *$+B^{m=2}$*.

<u>Example 3.9.</u> Binary signs the local structure models of second degree binary graphs *$g_{ij}{}^{m=2}$* of *Sib_A* and *Sib_B* correspondingly:

    Binary signs of second degree binary graph *$g^{m=2} \subset Sib_A$* in local structure model *$SM_{ij}{}^{m=2}{}_A$*:
    –A=–2.6.8; –B=–2.4.4; –C=–2.3.2; D=+2.4.6; E=+3.12.28; F=+3.20.46.

    Binary signs of second degree binary graph *$g^{m=2} \subset Sib_B$* in local structure model *$SM_{ij}{}^{m=2}{}_B$*:
    –A=–2.6.8; –B=–2.4.4; C+2.4.6; D=+3.12.24; E=+3.20.46.

Explanations: **a)** From differences of binary signs conclude *non-isomorphism* of second degree binary graphs. **b)** From non-isomorphism the binary graphs conclude non-isomorphism of graphs *Sib_A* and *Sib_B*.

Proposition 3.3. From non-isomorphism the binary graphs *$g_{ij}{}^A$* and *$g_{ij}{}^B$* of corresponding symmetric graphs *G_A* and *G_B* conclude ***non-isomorphism*** of *G_A* and *G_B*.

For illustrating the differences of *Sib_A* and *Sib_B* is suitable demonstrate second degree binary graphs.

<u>Example 3.10.</u> The kernels of second degree binary graphs *$g^{m=2}{}_A$* and *$g^{m=2}{}_B$* of very similar structures *Sib_A* and *Sib_B*:

Kernel of *$g_{1-6}{}^{m=2} \subset Sib_A$*:

Kernel of *$g_{20-22}{}^{m=2} \subset Sib_B$*:

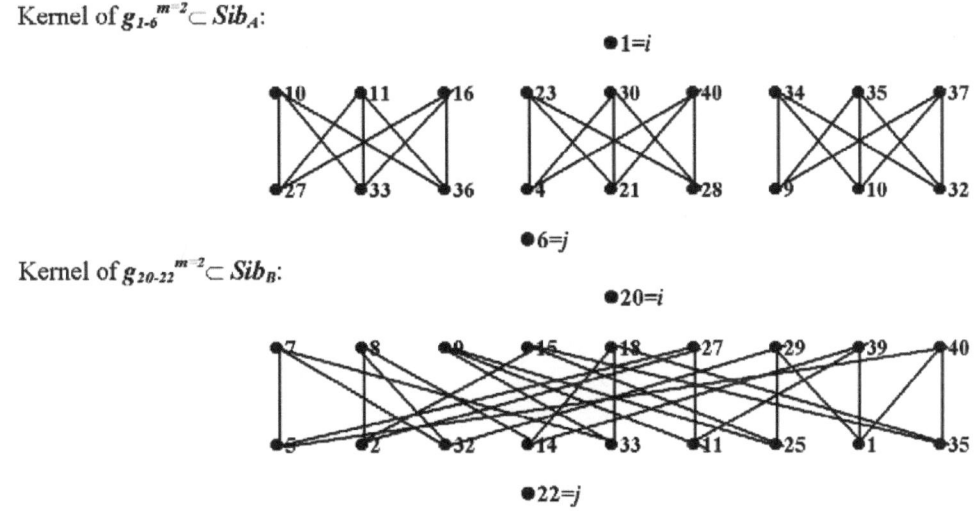

3.4. Outputs of two isomorphism algorithms

Only few isomorphism recognition algorithms give a canonical output of processing results. Usually be limited laconically with phrase "isomorphic" or "not isomorphic". We show the **canonical output** of two algorithms the isomorphism recognition. Recognition the orbits no belong to isomorphism problem. In historical journey so far, the question remains: *is the graph isomorphism problem in **P**?*

First, Dharvadker-Tevet polynomial algorithm [6] based on *incomplete semiotic models* S_A and S_B, where the vertex classes V_{Ak} and V_{Bk} given on the level only of frequency vectors. Notable here are the following moments: 1) Isomorphism recognition of this algorithm and its **polynomial** time complexity are proved in detail. 2) Transposition the rows *i* and columns *j* take place within vertex classes of corresponding partial models S_{Ak} and S_{Bk} and leaded to isomorphism recognition with **exactness up to substitutions**.

Example 3.11. Canonical outputs of Dharwadker-Tevet isomorphism algorithm:

Incomplete structure model of G_B:

Matrix B	1	8	7	3	6	4	5	2
1	-0.1.0	-2.8.21	+2.5.7	+2.5.7	+2.5.7	+2.5.7	+2.5.7	+2.5.7
8	-2.8.21	-0.1.0	+2.5.7	+2.5.7	+2.5.7	+2.5.7	+2.5.7	+2.5.7
7	+2.5.7	+2.5.7	-0.1.0	-2.7.16	-2.7.16	+2.4.5	+2.4.5	+2.4.5
3	+2.5.7	+2.5.7	-2.7.16	-0.1.0	-2.7.16	+2.4.5	+2.4.5	+2.4.5
6	+2.5.7	+2.5.7	-2.7.16	-2.7.16	-0.1.0	+2.4.5	+2.4.5	+2.4.5
4	+2.5.7	+2.5.7	+2.4.5	+2.4.5	+2.4.5	-0.1.0	-2.7.16	-2.7.16
5	+2.5.7	+2.5.7	+2.4.5	+2.4.5	+2.4.5	-2.7.16	-0.1.0	-2.7.16
2	+2.5.7	+2.5.7	+2.4.5	+2.4.5	+2.4.5	-2.7.16	-2.7.16	-0.1.0

Substitutions of graphs G_A and G_B:

Graph G_A	Graph G_B
4	1
5	8
1	7
2	3
3	6
7	4
8	5
6	2

Isomorphic graphs G_A and G_B;

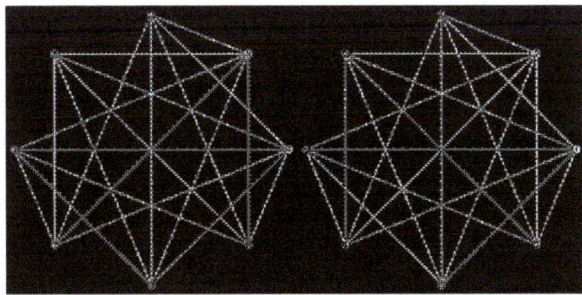

Explanations: a) In this example can be the semiotic models S_A and S_B treated as complete, because for their decomposition be suffice the exactness of frequency vectors. c) By this algorithm be recognized isomorphism also of canonically hardly recognizable graphs, for example, the non-isomorphism of *strongly symmetric* graphs Sib_A and Sib_B (see Example 3.10).

*

To canonical output of a graph in isomorphism algorithm of Blazej Podsiadlo [22] is its *biggest value* that no contain data about the graph, but enable to differentiate these, better as for example 3-cube-codes. It do no realized up to substitutions. To the canonical output belong *the biggest value* <**the biggest value**>, *the number of paths* <**paths**>, *the number of automorphisms* <**automorphisms**>, *the real time* <**treal**>.

Examples 3.12. A result of isomorphism algorithm, by Blazej Podsiadlo:

<**example**> <**paths**> <**automorphisms**> <**treal**> <**the biggest value**>

<**1A**> <**720**> <**79**> <**0m0.074s**>
1250779006614106429870497093178760201282717954568043451146532584283558626000476535653
12083251821991363786070369590522710 0

<**1B**> <**720**> <**34**> <**0m0.074s**>
1250779006614106429870497093178760201282717954568043451146532584283558626000476535621
21928075739411838235109474710637810 0
Result: **NOT Isomorphic**

Explanations: a) It ensues on the rather great coincidence in the beginning of sums **83/121** or **68,60%** similar. b) It is a performance with two vertex symmetric graphs Pra_A and Pra_B (Example 3.8) that have common first degree pair signs and are very similar.

The "length" of value depends on the vertex number and coincidence on relation the "lengths" of intersection and full value. In original program comparison the sums do not exist. As I have experience with these graphs, the results seem logical and acceptable. Naturally, their essence needs to research.

4. STRUCTURAL TRANSFORMATIONS

Structural transformations take place in case *transition* from a graph **G** to another graph **H**. *Elementary structural transformation* is a transition from graph **G** to its subgraph **G\v_i** or **G\e_{ij}**. To the relationships between graphs and its subgraphs was interested beginning at the formulation of Ulam Conjecture [44].

Suitable to remind a theorem about the relationships between **G** and its subgraphs **G\v_i**, proved by V. Titov in 1975 [41], which at that time, unfortunately, has not found the attention.

Titov's theorem. If all the (**G\v_i**)-sub-graphs of graph **G** are isomorphic, then automorphism group *AutG* is transitive on the set of vertices **V**.

It mean that graph **G** is *transitive* or *vertex symmetric*, i.e. there exists only one vertex position $\Omega V_{k=1=K} = \Omega(v_{i=1}, ..., v_{i=|V|})_{k=1=K}$, which correspond just to one isomorphism class of (**G\v_i**)-sub-graphs, $\Gamma_{k=1=K} = (G \backslash v_{i=1} \cong ... \cong G \backslash v_{i=|V|})_{k=1=K}$.

4.1. Structural transformations and reconstructions

By removing an edge **G\e_{ij}** of **G** obtained a *greatest subgraph* **G^{sub}**. The number of **G^{sub}** equals to the number of edges. With adding an edge **G∪e_{ij}** to **G** obtained a *smallest supergraph* **G^{sup}**. The number of **G^{sup}** equals to the number of "non-edges".

Definition 4.1. Greatest subgraphs **G^{sub}** and smallest supergraphs **G^{sup}** called *adjacent graphs* **G^{adj}** of **G**.

Proposition 4.1. If the adjacent graphs **G^{adj}** are obtained on the ground of the same binary position ΩR_n then are these *isomorphic* and constitute an *adjacent structure* **GS^{adj}_n** of **GS**. The number of adjacent structure equals to the number of binary positions.

Corollary 4.1. Disjunctive edge operation $F_n = \{(f_{ij})_1 \vee ... \vee (f_{ij})_q\}_n$ in the frame of a binary position ΩR_n that *transforms* the structure **GS** to its adjacent structure **GS^{adj}_n** is called *morphism*, $F_n: GS \rightarrow GS^{adj}_n$.

Example 4.1. *Partially symmetric* structure *GS.37(6.9.4)* (see Supplement) with two element positions and four binary positions, its graph, structure model, characteristics of transformations and morphisms:

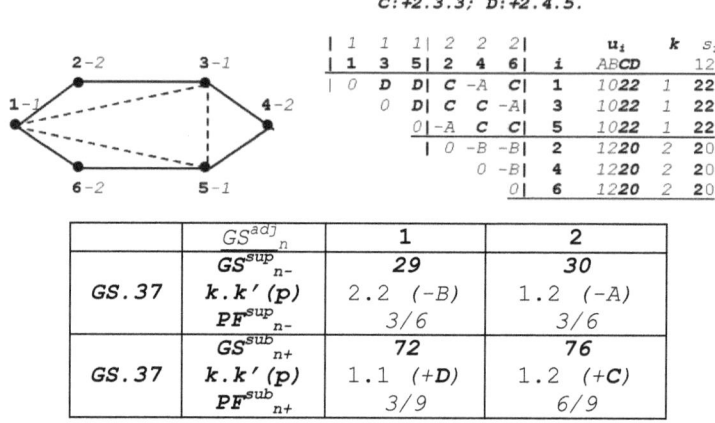

	GS^{adj}_n	1	2
	GS^{sup}_{n-}	29	30
GS.37	k.k'(p)	2.2 (−B)	1.2 (−A)
	PF^{sup}_{n-}	3/6	3/6
	GS^{sub}_{n+}	72	76
GS.37	k.k'(p)	1.1 (+D)	1.2 (+C)
	PF^{sub}_{n+}	3/9	6/9

Explanations:
  a) GS^{sup}_{n-} and GS^{sub}_{n+} denotes the *ordering numbers* of adjacent superstructures and adjacent substructures in the system of structures with six elements (example 4.4);
  b) ***k,k'*** – index of partial model $SM_{k,k'}$, whither belong the binary position *(p)*;
  c) PF_n – *morphism probability*.

Example 4.2. Three *isomorphic graphs* that represent the ***adjacent superstructure*** $GS^{sup}_{n=-B}$, *(GS.29)* (Supplement) of structure *GS.37* (example 4.1). These are obtained by *adding* the connections 2-4 or 2-6 or 4-6 (dashed line) to binary(−)position −B of GS.37. Their *common binary signs* and *equivalent models* **SM₁ ≡ SM₂ ≡ SM₃**:

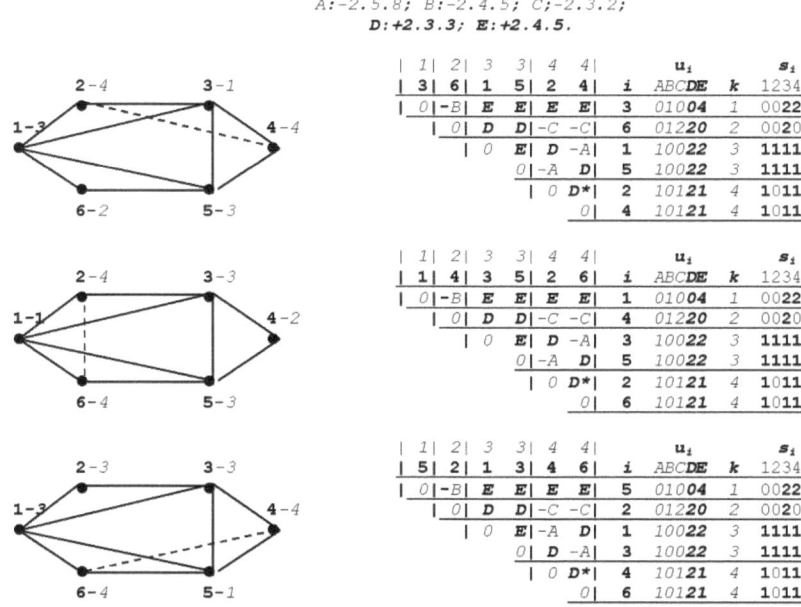

```
A:-2.5.8; B:-2.4.5; C:-2.3.2;
     D:+2.3.3; E:+2.4.5.
```

1	2	3	3	4	4		u_i		s_i
3	6	1	5	2	4	i	ABCDE	k	1234
0	-B	E	E	E	E	3	01004	1	0022
	0	D	D	-C	-C	6	01220	2	0020
		0	E	D	-A	1	10022	3	1111
			0	-A	D	5	10022	3	1111
				0	D*	2	10121	4	1011
					0	4	10121	4	1011

1	2	3	3	4	4		u_i		s_i
1	4	3	5	2	6	i	ABCDE	k	1234
0	-B	E	E	E	E	1	01004	1	0022
	0	D	D	-C	-C	4	01220	2	0020
		0	E	D	-A	3	10022	3	1111
			0	-A	D	5	10022	3	1111
				0	D*	2	10121	4	1011
					0	6	10121	4	1011

1	2	3	3	4	4		u_i		s_i
5	2	1	3	4	6	i	ABCDE	k	1234
0	-B	E	E	E	E	5	01004	1	0022
	0	D	D	-C	-C	2	01220	2	0020
		0	E	-A	D	1	10022	3	1111
			0	D	-A	3	10022	3	1111
				0	D*	4	10121	4	1011
					0	6	10121	4	1011

Explanation: Equivalent models differ from each other only by numbering of elements in the positions.

Example 4.3. The *different* ***adjacent substructures*** $GS^{sub}_{n=+D}$, *(GS.72)* (Supplement) and $GS^{sub}_{n=+C}$, *(GS.76)* of structure *GS.37* (example 4.1) that obtained by *removing* the connection 3-5 from binary(+)position +D and *removing* the connection 5-6 from binary(+)position +C correspondingly. Their *non-isomorphic graphs, different binary signs* and *non-equivalent structure models* **SM_A** and **SM_B**:

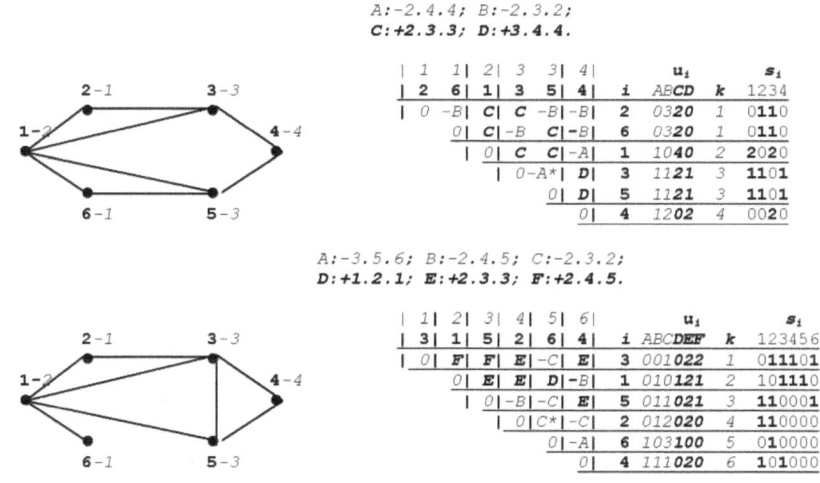

```
A:-2.4.4; B:-2.3.2;
  C:+2.3.3; D:+3.4.4.
```

1	1	2	3	3	4		u_i		s_i
2	6	1	3	5	4	i	ABCD	k	1234
0	-B	C	C	-B	-B	2	0320	1	0110
	0	C	-B	C	-B	6	0320	1	0110
		0	C	C	-A	1	1040	2	2020
			0	-A*	D	3	1121	3	1101
				0	D	5	1121	3	1101
					0	4	1202	4	0020

```
A:-3.5.6; B:-2.4.5; C:-2.3.2;
  D:+1.2.1; E:+2.3.3; F:+2.4.5.
```

1	2	3	4	5	6		u_i		s_i
3	1	5	2	6	4	i	ABCDEF	k	123456
0	F	F	E	-C	E	3	001022	1	011101
	0	E	E	D	-B	1	010121	2	101110
		0	-B	-C	E	5	011021	3	110001
			0	C*	-C	2	012020	4	110000
				0	-A	6	103100	5	010000
					0	4	111020	6	101000

43

Explanation: On the basis of various binary positions obtained adjacent structures are *not equivalent*.

Each structure GS is an adjacent substructure GS^{sub}_n or adjacent superstructure G^{sup}_n of some other structures. For each binary position ΩR_n correspond an adjacent structure GS^{adj}_n.

Proposition 4.2. If morphisms F_n: $GS \rightarrow GS^{adj}_n$ are applied to binary positions $\Omega R_1,...,\Omega R_n,...,\Omega R_N$ of GS disjunctively, $F_1 \vee ... \vee F_n \vee ... \vee F_N$, then GS is **transformed** to its *adjacent structures* $GS^{adj}_1,...,GS^{adj}_n,...,GS^{adj}_N$.

Corollary 4.2. *Not-transformable structures do not exist.*

Proposition 4.3. Morphism F is **inverses** – in each adjacent structure GS^{adj} of GS exist an "*inverse position*" ΩR^{inv}, whereat used *inverse morphism* F^{inv} reconstruct the initial structure GS, F^{inv}: $GS^{adj} \rightarrow GS$.

The *morphism probability* PF_n of the structure $GS.37$ on example 4.1 to the adjacent structure $GS.29$ is $3/6$. If $GS.29$ is an initial structure GS then the *morphism probability of reconstruction* of its adjacent substructure GS^{sub}_n is $PF^{inv} = 1/10$. Therefore, the morphism probability PF_n depends from the *initial structure* GS.

Propositions 4.4. The relations between transformations and reconstructions of structure:
P4.4.1. If structure GS is **transformed** to its *adjacent substructures* $GS^{sub}_1,...,GS^{sub}_n,...,GS^{sub}_N$, then GS is the *common adjacent superstructure* GS^{sup} of all its adjacent substructures GS^{sub}_n and is *reconstructable* by an *inverse morphism* F^{inv}_n: $GS^{sub}_n \rightarrow GS$ to each its adjacent substructure.
P4.4.2. If structure GS is **transformed** to its *adjacent superstructures* $GS^{sup}_1,...,GS^{sup}_n,...,GS^{sup}_N$, then GS is the *common adjacent substructure* GS^{sub} of all its adjacent superstructures GS^{sup}_n and is *reconstructable* by an *inverse morphism* F^{inv}_n: $GS^{sup}_n \rightarrow GS$ to each its adjacent superstructure.

Thus, structure GS is **reconstructable** by its *adjacent substructures* GS^{sub}_n and its *adjacent superstructures* GS^{sup}_n. This coexistence makes the reconstruction to an **inverse transformation**.

Corollary 4.3. *Not-reconstructive structures do not exist.*

The *reconstruction problem* is known as **Ulam's Conjecture** that reflects the isomorphism relations between two graphs and their $(G\backslash v_i)$-subgraphs [44]. It is formulated as follows: "If for each i, the subgraphs $G_i = G\backslash v_i$ and $H_i = H\backslash u_i$ are isomorphic, then the graphs G and H are isomorphic".

This problem has been over the past half century, one of under active consideration graph theoretical problem, but the ultimate solutions have only some graph classes. Why so? On the structural aspect are the attempts of solution the conjecture by its wording *senseless,* because, if given graphs G and H then on the ground of *structure models* SM_G and SM_H we obtain the complete information about corresponding graphs, their isomorphism or non-isomorphism and of their adjacent graphs. Other approaches are meaningless for us here.

Evidently be interested on the question: contains the collection of subgraphs $G\backslash v_i$ of G enough information about graph G itself? Ulam's Conjecture treats the reconstruction on the aspect of removing of the vertices, but we treat it on the aspect of adding and removing of edges. This not changes the essence of reconstruction, because all remains to the frame of graphs (structures) and their adjacent-graphs (adjacent-structures), i.e. in our case to the frame of *morphisms F_n*. Already old master W. T. Tutte emphasized that reconstruction-problem must be solve on the basis of *isomorphism classes, (i.e. structures)* that we also have followed [43].

4.2. Systems of structural transformations

Each set of all the non-isomorphic graphs with **n** vertices constitute a *system of adjacent structures* i.e. *system of structural transformations*. Each structure can to its adjacent structures transformed and each structure is an adjacent structure of some structures.

By help of morphisms F_n are generated the *system of adjacent structures with five elements* [36] (where 72 morphisms connect 34 structures) and the *system with six elements* [37] (where 572 morphisms connect 156 structures). It can be generated for all tructures and shows the inevitability of reconstructing.

Example 4.4. The *lattice of transformations of the structures* with six elements (see also Supplement):

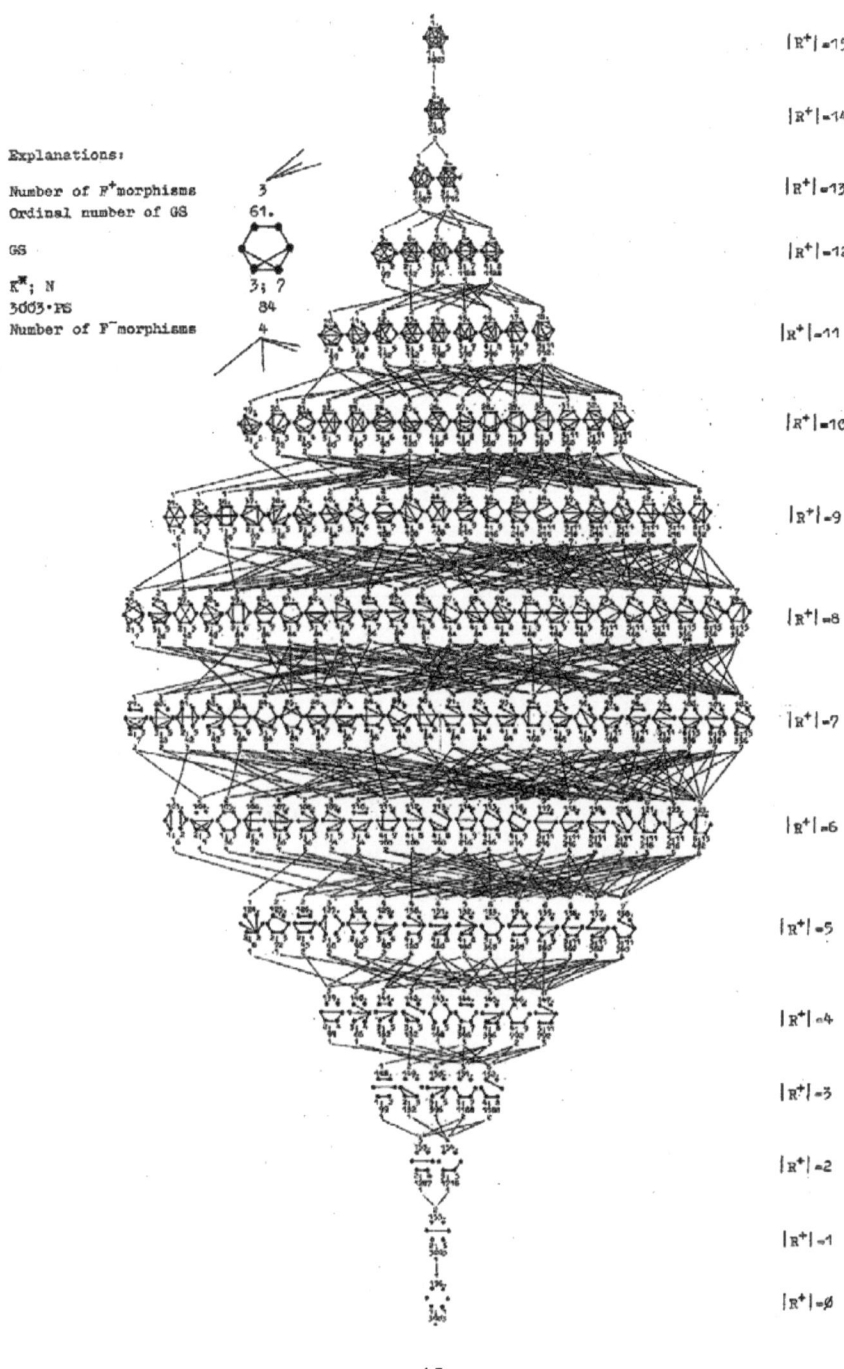

45

Explanations:
   a) $|\mathbf{R}^+|$ denote the *structural level m*, i.e. the number of connections (i.e. "edges") in the structures.
   b) Each graph presents there its *isomorphism class* or *structure GS*.
   c) Each structure in this lattice is an *adjacent structure* of some other structures, where the edges represent the morphisms F_n.
   d) The *complements* of represented structures placed symmetrically in the upper and lower half of lattice.

Generating of the adjacent structures proceed by the morphisms F_n, so that in the framework of structures *GS* of a structural level *m* (i.e. structures with a concrete number of "edges") be formed the structure models of adjacent structures GS^{adj}. And so proceed, from a level to its adjacent level [33]. In result is obtained the *system of structural transformations* or *system of adjacent structures* ⑤. Generating can be begin from zero or complete structure (see also Supplement).

Formed many collections of graphs with *n* vertices, but the adjacencies do not notice. Why is that? The first sample of non-isomorphic graphs with up to six vertices was represented by Frank Harary in 1969th [10]. Later, F. Harary and E. Palmer had calculated the number of non-isomorphic graphs (i.e. structures) up to 24 vertices [11]. R. Read and F. Wilson have given the diagrams of graphs also up to seven vertices [24]. But so far about the relationships between structures, i.e. morphisms, do not discussed – their do not wanted to notice.

Example 4.5. The number of graphs in the samples of non-isomorphic graphs with 3 to 10 vertices:

Number of elements $|V|$	Number of structures p	Among this connected p^*	Number of levels m	Number of morphisms q
3	4	2	4	3
4	11	6	7	14
5	34	21	11	72
6	156	112	16	572
7	1044	853	22	
8	12346	11117	29	
9	274668	261080	37	
10	12005168	11716571	46	

Propositions 4.5. Some *general properties* of the systems $⑤^{|V|}$:
P4.5.1. If the number of structural levels *m* in system $⑤^{|V|}$ is *even number* (as in case $|V|=6$ and $|V|=7$), then it lattice is *bilaterally symmetric* with regard its bisector, which separates the *structures GS* from their *complements* ⌉*GS*.
P4.5.2. If the number of structural levels *m* in system $⑤^{|V|}$ is *odd number* (as in case $|V|=4$, $|V|=5$, $|V|=8$ and $|V|=9$), then the bisector is a structural level in which be located the *structures GS*, their *complements* ⌉*GS* and also *self-complemented* structures *GS*= ⌉*GS*.

In structural genesis has important role *randomness*. This is expressed in the form of *selection the adjacent structures*, i.e. elementary structural changes. The *probabilistic characteristics* are related with *internal diversity* of structure, i.e. binary positions, and have essential meaning in structural research.

Propositions 4.6. *Probabilistic characteristics* of the systems $⑤^{|V|}$:
P4.6.1. *Randomness* in the systems ⑤ based on the *morphism probabilities PF_n*.
P4.6.2. There exists *transition probability P_{ij}* at a structure GS_i to a non-adjacent structure GS_j.
P4.6.3. Transition probabilities P_{ij} form the *stationary Markov chain PM* of structural genesis.

P4.6.4. *Existence probability* **PS** of structure **GS** in the structural level $|\mathbf{R}^+|$ of system $\mathfrak{G}$ is expressed in the form:

$$PS =_{n=1}\sum^{N-} PS^{sup}_{n} \times PF^{sub}_{n} =_{n=1}\sum^{N+} PS^{sub}_{n-} \times PF^{supp}_{n,}$$

where **n** is the structural index of binary position, PS^{sup}_{n} existence probability of adjacent superstructure and PF^{sub}_{n} its morphism probability.

P4.6.5. The *sum of existence probabilities* **PS** of structures in the structural level $|\mathbf{R}^+|$ equal to one, $\sum PS = 1$.

P4.6.6. Existence probabilities of *structure* and its *complement* are equal, **PS(GS)=PS(⌐GS)**.

P4.6.7. Existence probabilities **PS** are *rational numbers* and directly related with the degree of genesis $|V|$.

P4.6.8. Distribution of the probabilities **PS** in the structure levels approach to *logarithmic normal distribution*.

Since morphisms have by structural transformations principal role, they are suitable to represent any propositions about them. Some algebraic systems can express some fragments or aspects of structural transformations and can have corresponding models.

Proposition 4.7. The class of morphisms **F** forms an ***additive group A*** from the aspect of the compositions ***F&F*** in the system $\mathfrak{G}^{|V|}$.

Let us check the validity of additive group postulates for morphisms:
1) $F_a \& (F_b \& F_c) = (F_a \& F_b) \& F_c$ (distributivity).
2) $\forall (F_a, F_b)$; $F_a, F_b \in A$, $\exists F_c$, $F_a \& F_b = F_c \in A$ (because composition of morphisms is also a morphism).
3) $\forall F \exists! F'$; $F, F' \in A$, $\forall (F, F') \exists F \& F' = F^0 \in A$ (existence of an opposite morphism).
4) $\exists! F^0$; $F^0 \in A$, $\forall (F, F^0) = F \in A$ (existence of a zero morphism).

The postulates of the *category C* satisfy the system $\mathfrak{G}^{|V|}$ most. S.Eilenberg and S.MacLane [7] formulated the foundations of category **C** in 50th years in the framework of elaboration of algebraic methods for topology. The notion "morphism" is a principal concept of category **C**.

Proposition 4.8. The class of **G**-structures $\{GS^{|V|}\}$ together with morphism's class **F** of the system $\mathfrak{G}^{|V|}$ forms a ***category C***.

There exist correspondences of postulates of category with attributes of the system $\mathfrak{G}^{|V|}$. Let the class $\{GS^{|V|}\}$ correspond to object class **Ob** of **C**.
1) $\forall (GS_i, GS_j) \in \{GS^{|V|}\}$, $\exists \{F\}$: $(GS_i \rightarrow GS_j)$, that represents a set of possible morphisms from GS_i to GS_j, denoted by $\{F\} \subset \text{Hom}(GS_i, GS_j)$, which can be constitute various successions of structures **GS**.
2) For each $(GS_i, GS_j, GS_k) \in \{GS^{|V|}\}$ exist mapping $\text{Hom}(GS_i, GS_j) \& \text{Hom}(GS_j, GS_k) \rightarrow \rightarrow \text{Hom}(GS_i, GS_k)$, since $F_a \in \text{Hom}(GS_i, GS_j)$ and $F_b \in \text{Hom}(GS_j, GS_k)$, where the result of composition of morphisms (or succession, as an high degree morphism) $F_a \& F_b = F_c \in \text{Hom}(GS_i, GS_k)$.
3) Morphisms $\text{Hom}(GS_i, GS_j)$ and their compositions satisfy the category postulates, because:
   - In the case of each succession $^{Fa:}GS_i \rightarrow^{Fb:} GS_j \rightarrow^{Fc:} GS_k \rightarrow$ the associability $F_a \& (F_b \& F_c) = (F_a \& F_b) \& F_c$ be valid;
   - $\forall GS \in \{GS^{|V|}\} \exists F^0 = F \& F'$; where F^0: $GS \rightarrow GS$ constitute identity morphism or an unit of **GS**;
   - If the pairs (GS_i, GS_j) and (GS_i', GS_j') are different, then $\text{Hom}(GS_i, GS_j) \cap \text{Hom}(GS_i', GS_j') = \varnothing$ which mean the existence of disjoint connecting graphs $\mathfrak{G}^{|V|}_{ij} \cap \mathfrak{G}^{|V|}_{ij}' = \varnothing$ in the system $\mathfrak{G}^{|V|}$.

4.3. Successions of structural transformations

The interest for succession of structures, i. e. *structural transformations* is connected not only with the study of their lawfulness. The structural transformations have also been essential in cognitive and applicative analysis, research and simulation of phenomena and processes where the existence can be expressed as the gradual changes of structure [8, 16, 40].

Successive elementary transformation of structures **GS** can be expressed as the paths in the lattice of system $\mathfrak{G}^{|\mathcal{V}|}$ or can be formed independently.

Definition 4.2. An ordered set of morphisms $F_1 \& F_2 \& ... \& F_t$ to structures **GS**,
$$^{F1:}GS_0 \to {}^{F2:}GS_1 \to {}^{F3:}GS_2 \to ... {}^{Ft:}GS_{t-1} \to GS_t,$$
is a *succession of structures*, denoted by **SF**.

Propositions 4.9. The properties of *successions SF*:
P4.9.1. A succession **SF** can proceed **randomly** or **non-randomly**. If the selection of morphisms takes place on the ground of certain conditions or criterions, then it is a **teleological succession**.
P4.9.2. The successions between non-adjacent structures GS_i and GS_j, in the lattice of the system $\mathfrak{G}^{|\mathcal{V}|}$ constitutes an **assemblage of successions**.
P4.9.3. Structural changes that take place only by F^+- or only F^- morphisms form a **vertical succession**.
P4.9.4. A succession **SF** whose initial structure GS_i and result structure GS_j can be found on the same subsystem $\mathfrak{G}^m$ form a **horizontal succession**. Such structural changes are based on the morphism's pair $F^- \& F^+$ (or $F^+ \& F^-$), which constitutes a *"displacement of connection"* in **GS**.
P4.9.5. A succession **SF** where the structural values change monotonously, constitutes **monotonous succession SFM** on the sense of corresponding characteristics.
P4.9.6. *Probability of random succession PSF* with length t constitutes the product of corresponding morphism probabilities,
$$PSF = {}_{i=1}\prod{}^t PF_i = PF_1 \times PF_2 \times ... \times PF_t.$$

In $\mathfrak{G}^{|\mathcal{V}|=6}$ can be exist *the assemblages of horizontal successions* between two structurally extremely different **GS**. For example, in the subsystem $\mathfrak{G}^{m=9}$ between the symmetrical structure GS_{34} *(SR=0.751)* and totally asymmetric structure GS_{54} *(SR=0)*. Let us fix the partial system $\mathfrak{G}_{34;54}$ as an assemblage of horizontal successions that include the structures **GS** from neighbouring subsystems $\mathfrak{G}^{m=8}$ and $\mathfrak{G}^{m=10}$. The length of successions $d=4$ (with only two "displacements"), the number of structures in $\mathfrak{G}_{34;54}$ $n=12$ and the number of morphisms $q=22$. A single succession of the assemblage $GS_{34} \to GS_{57} \to GS_{52} \to GS_{78} \to GS_{54}$, is monotonous. This example is essential for the study of random aspects of structural changes.

Example 4.7. Partial system $\mathfrak{G}^{|\mathcal{V}|=6}_{34,54}$ representing an *assemblage of horizontal successions*:

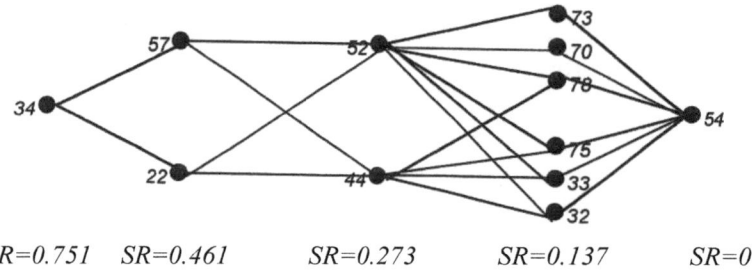

$SR=0.751 \quad SR=0.461 \quad SR=0.273 \quad SR=0.137 \quad SR=0$

Explanation: Symmetry values are represented by an upper monotonous succession.

The probabilities of random successions in assemblage of horizontal successions in subsystem $\mathfrak{G}^{|V|=6}_{34,54}$ are products of random morphisms. Probability of changing a symmetric structure GS_{34} to a completely **a**symmetric structure GS_{54} is $PSF_{C1}=9/15\times4/15\times2/15\times1/15=72/50625$. The same succession, but in contrary direction – changes completely **a**symmetric structure GS_{54} to a symmetric structure GS_{34} with probability $PSF_{C2}=1/15^4=1/50625$. Thus probability of changing an **a**symmetric structure GS to a symmetric *is 72 times less* than in opposite case! It is rather a rule than a chance.

<u>Example 4.8.</u> *Stationary Markov chain* $PM_{34,54}$ for assemblage of horizontal successions in subsystem $\mathfrak{G}^{|V|=6}_{34,54}$:

GS	34	22	57	44	52	32	33	75	78	70	73	54
34	0	$\frac{1}{20250}$	$\frac{1}{30375}$	$\frac{2}{8100}$	$\frac{2}{16200}$	$\frac{3}{2160}$	$\frac{3}{2160}$	$\frac{3}{2160}$	$\frac{3}{4320}$	$\frac{3}{2160}$	$\frac{3}{2160}$	$\frac{4}{1728}$
22	$\frac{1}{3375}$	0	$\frac{2}{4050}$	$\frac{1}{10125}$	$\frac{2}{20250}$	$\frac{2}{2700}$	$\frac{2}{2700}$	$\frac{2}{2700}$	$\frac{2}{5400}$	$\frac{2}{2700}$	$\frac{2}{2700}$	$\frac{3}{2160}$
57	$\frac{1}{3375}$	$\frac{2}{2700}$	0	$\frac{1}{6750}$	$\frac{1}{13500}$	$\frac{2}{1800}$	$\frac{2}{1800}$	$\frac{2}{1800}$	$\frac{2}{3600}$	$\frac{2}{1800}$	$\frac{2}{1800}$	$\frac{3}{1440}$
44	$\frac{2}{450}$	$\frac{1}{3375}$	$\frac{1}{3375}$	0	$\frac{2}{4725}$	$\frac{1}{6750}$	$\frac{1}{6750}$	$\frac{1}{6750}$	$\frac{1}{13500}$	$\frac{3}{405}$	$\frac{3}{405}$	$\frac{2}{3600}$
52	$\frac{2}{350}$	$\frac{1}{3375}$	$\frac{1}{3375}$	$\frac{2}{1800}$	0	$\frac{1}{3375}$	$\frac{1}{3375}$	$\frac{1}{3375}$	$\frac{1}{6750}$	$\frac{1}{6750}$	$\frac{1}{6750}$	$\frac{2}{3600}$
32	$\frac{3}{60}$	$\frac{2}{450}$	$\frac{2}{450}$	$\frac{1}{3375}$	$\frac{1}{3375}$	0	$\frac{2}{1125}$	$\frac{2}{1125}$	$\frac{2}{2025}$	$\frac{2}{2700}$	$\frac{2}{2700}$	$\frac{1}{6750}$
33	$\frac{3}{60}$	$\frac{2}{450}$	$\frac{2}{450}$	$\frac{1}{3375}$	$\frac{1}{3375}$	$\frac{2}{1125}$	0	$\frac{2}{1125}$	$\frac{2}{2025}$	$\frac{2}{2700}$	$\frac{2}{2700}$	$\frac{1}{6750}$
75	$\frac{3}{60}$	$\frac{2}{450}$	$\frac{2}{450}$	$\frac{1}{3375}$	$\frac{1}{3375}$	$\frac{2}{1125}$	$\frac{2}{1125}$	0	$\frac{2}{2025}$	$\frac{2}{2700}$	$\frac{2}{2700}$	$\frac{1}{6750}$
78	$\frac{3}{60}$	$\frac{2}{450}$	$\frac{2}{450}$	$\frac{1}{3375}$	$\frac{1}{3375}$	$\frac{2}{1125}$	$\frac{2}{1125}$	$\frac{2}{1125}$	0	$\frac{2}{2700}$	$\frac{2}{2700}$	$\frac{1}{3375}$
70	$\frac{3}{60}$	$\frac{2}{450}$	$\frac{2}{450}$	$\frac{3}{420}$	$\frac{1}{6750}$	$\frac{2}{2700}$	$\frac{2}{2700}$	$\frac{2}{2700}$	$\frac{2}{1350}$	0	$\frac{2}{1350}$	$\frac{1}{6750}$
73	$\frac{3}{60}$	$\frac{2}{450}$	$\frac{2}{450}$	$\frac{3}{420}$	$\frac{1}{6750}$	$\frac{2}{2700}$	$\frac{2}{2700}$	$\frac{2}{2700}$	$\frac{2}{1350}$	$\frac{2}{1350}$	0	$\frac{1}{6750}$
54	$\frac{4}{24}$	$\frac{3}{180}$	$\frac{3}{180}$	$\frac{2}{900}$	$\frac{2}{1800}$	$\frac{1}{3375}$	$\frac{1}{3375}$	$\frac{1}{3375}$	$\frac{1}{3375}$	$\frac{1}{3375}$	$\frac{1}{3375}$	0

Explanations:
 a) The numbers 1 to 4 represent the number of steps or distance d.
 b) The numbers 24 to 30375 represent transition probabilities P_{ij} multiplied 50625 times.
 c) We see that transition probability from symmetric structure GS_{34} to asymmetric structure GS_{54} $P_{34,54}=1728{:}50625$ *is* $1728{:}24=72$ *times less* than in opposite case, $P_{54,34}=24{:}50625$!

Asymmetric structures dominate. On the structural aspect are "asymmetrical" the *0-symmetric* and *partially symmetric* structures. In the system $\mathfrak{G}^{|V|=6}$ are eight 0-symmertic (5.73%), 140 partially symmetric structures (89.7%) and eight symmetric (edge- and polysymmetric) structures. Relative occurrence frequency of "asymmetric" structures, notably of 0-symmetric structures, *increase* successfully by enlargement of degree $|V|$.

4.4. Monotonous structural successions as dynamic or evolutional phenomena

Structure constitutes something *qualitative*. Thus, each possible structural transformation is a *change of quality*. At the same time is structure *measurable* from various aspects. Each separate measure, i.e. structural characteristic *H*, expresses evidently a single quantitative aspect, but with a certain set of structural characteristics it is possible to recognize also qualitative properties of the structure.

Now we can speak only on differences on the aspect of their structural characteristics. Does such *vertical succession*, which pervade all the structural levels so, that its *structural characteristics change monotonously?* It is ascertained, that for the great number of various vertical successions in $\mathfrak{G}^{|V|=6}$ there exists only one single succession, which satisfies such strict conditions completely $GS_1 \rightarrow GS_2 \rightarrow GS_4 \rightarrow GS_8 \rightarrow GS_{17} \rightarrow GS_{33} \rightarrow GS_{53} \rightarrow GS_{76} \rightarrow GS_{100} \rightarrow GS_{122} \rightarrow GS_{138} \rightarrow GS_{146} \rightarrow GS_{151} \rightarrow GS_{154} \rightarrow GS_{155} \rightarrow GS_{156}$. Such structural characteristics are compactness *CMP*, existence probability *PS*, symmetry value *SR* (or contrary –asymmetricality *CR*), triangularity *TRA*, branching *FRK*, complexity *CPX*, topological entropy *HE*, diameter *DMR*, cliqueability *MCQ* and information capacity *HV*.

Such monotonous succession and corresponding structural characteristics was useable by constructing of an elegant but very abstract ontogeny phenomenon – at origin to ripeness [29].

Example 4.9. A diagram of the *monotonous succession SFM* in the system $\mathfrak{G}^{|V|=6}$:

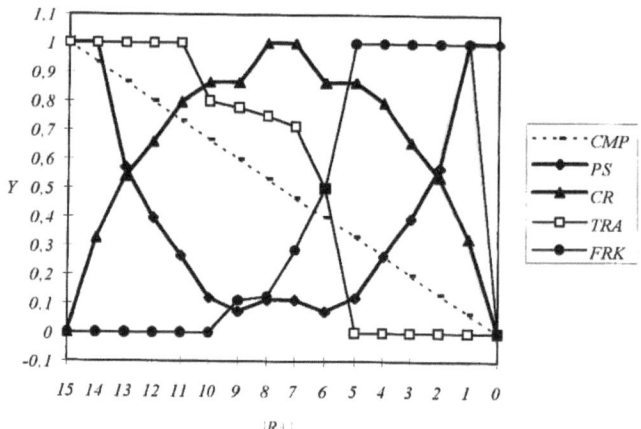

Explanation: a) *Y* – rationed value of structural characteristics; *R* – ordering number of structural level *GSL*. b) If the monotonous succession *SFM* is random then its probability PSF_B^- $=2/(4725 \times 3003)=1.41 \times 10^{-7}$ is of extremely little value.

There exist real systems where their functioning can be expressed by gradual changes of structure in time. If the structures *GS* of the system $\mathfrak{G}^{|V|}$ are treated as *states S of real system* then a such succession, $SF=$ $^{F1:}GS_0 \rightarrow^{F2:} GS_1 \rightarrow^{F3:} GS_2 \rightarrow... \rightarrow^{Ft:} GS_{t-1} \rightarrow GS_t$ constitutes a *dynamic or evolutional phenomenon*, generated by morphisms.
Let the addition- or elimination operations *{f}* be treated as certain *input impacts*. Let morphisms as the sets of disjunctive operations $F_n=\{f_1 \vee...\vee f_q\}_n$ form a *class of effects* **F**.

Let the values of structural and probabilistic characteristics form an *output class* **Y**. If the set $\{GS^{|V|}\}$ of all structures in the system $\mathfrak{G}^{|V|}$ be treated as a *state class* **S**, then the set **Y** of all the output values will represent a *phase space* of this system. An every time moment $t \in$ **T** the system receives an input impact $f_t \in$ **F**, which changes the state S_{t-1} as a current step of succession, to a next state S_t. Each input impact f_t belongs also to a certain *guide class* **X**, $f_t \in$ **F** $\subset$ **X**, which determines the possible change of a current state

S_{t-1}. The selection of the class **F** can be ***random,*** or on the contrary related with a ***functional objective Z*** of the system.

The current values $y_t \in$ **Y** of output set characterize a system state S_t, $y_t = \lambda(S_t)$. Knowing of the state S_t and fixing of an input impact $f_{t'} \in F_n$ is necessary and sufficient for determination of the state $S_{t'} = \varphi(S_t, f_{t'})$, always if $t < t'$.

It should also be noted, that our concept of ***dynamic system DYS*** is more concrete than the well-known concept of ***stationary (causal) system*** where each "present moment" ("past") changes to a "prospective moment" ("future"). In the case of dynamic system ***DYS***, it can be based on the system $\mathfrak{S}^{|I|}$, where:

  a)   one and the same "past" can be changed to various different "futures";
  b)   different (various) "pasts" can be changed to one and the same common "future".

Let us apply the concepts of the system of structural changes in $\mathfrak{S}^{|I|}$ to the classical concepts of the dynamic systems [14].

Propositions 4.10. Correspondence between attributes of the ***system*** $\mathfrak{S}^{|I|}$ and classical postulates of ***dynamic system DYS***:

P4.10.1. Correspondences of the sets:

  a)   The set $\{GS^{|I|}\}$ of all the structures in $\mathfrak{S}^{|I|}$ corresponds to a ***state class*** **S**;
  b)   the set $\{t\}$ of all the steps of successions corresponds to an ***ordered set of time moments (or stages)*** **T**;
  c)   the set $\{f\}$ of f-operators corresponds to a ***effects class*** **F**;
  d)   the effects class **F** belong to ***guide class*** **X**;
  e)   the set of structural and probabilistic characteristics corresponds to a ***output class*** **Y** of the system ***DYS***.

P4.10.2. There exists an ***output mapping*** λ: **T**×**S**→**Y** that determines the output values $y_t \in$ **Y** by each state S_t.

P4.10.3. The set of possible input impacts **X** be ***contracted*** to the classes of actual effects **F**.

P4.10.4. There exists an ***objective function*** μ: **T**×**Z**×**X**→**F**, such that at every time moment (stage) $t \in$ **T** on the basis of f-impacts $f \in$ **X** such a class $F_n \in$ **F** is selected, which could help reach a ***functional objective*** **Z** in the best possible way. Functional aim **Z** is realized in the form of an aim state ZS_j or behaviour criterion ΔZ (in the form of monotonous chains).

P4.10.5. The main attribute of dynamic system ***DYS*** is a ***succession function*** φ: **T**×**S**×**F**→**S**, where its values are states $S = \varphi(t'; t, S, f) \in$ **S**, in which the system ***DYS*** is at a time moment (stage) $t' \in$ **T**, if at the preceding time moment $t \in$ **T** it was in the preceding state $S \in$ **S** and has an effect of the input impact $f \in F_n \subset$ **X**. The function φ has the following classical characteristics:

  a)   ***direction of time proceeding:*** function φ is determined for every t, where $t < t' < t''$...;
  b)   ***a semi-group characteristic:*** by every $t < t' < t''$, every $S \in$ **S** and every $f \in$ **F** there   is $\varphi(t''; t, S, f) = \varphi(t''; t', \varphi(t'; t, S, f), f)$;
  c)   ***causality:*** if $f, f' \in F_n \subset$ **F** $\subset$ **X**, then $\varphi(t'; t, S, f) = \varphi(t'; t, S, f')$;
  d)   ***compatibility:*** equality $\varphi(t; t, S, f) = S$ is valid in the case of every $t \in$ **T**, every $S \in$ **S** and every $f \in$ **F**.

Now we can formulate some corollaries.

Corollaries 4.4. On the *successions SF* and discrete *dynamic system DYS:*

C4.4.1. Succession *SF* of structural changes represents a *dynamic or evolutional process.* Forming of succession constitutes a dynamic process itself.

C4.4.2. Discrete dynamic system *DYS* is *stationary,* if there exists a succession function φ. According to P4.10.4 the system *DYS* is *teleological,* if this does not hold, then *DYS* is *stochastic.*

C4.4.3. The set of system states **S** of *DYS* constitute a *factor space.*

C4.4.4. The various values of output characteristics **Y** of states form a *phase space* of *DYS.*

C4.4.5. The triplet (t,S,y), $t \in \mathbf{T}$, $S \in \mathfrak{G}^{|\Pi|}$, $y \in \mathbf{Y}$, represents an *event* and $\mathbf{T} \times \mathbf{S} \times \mathbf{Y}$ is an *event space* of *DYS.*

To the key problem of successions remain their steps as *elementary transformations* of structure: the difference between semiotic model **SM** of initial structure *GS* and semiotic model $\mathbf{SM}^{adj}$ of its adjacent structure GS^{adj} – the *extent of difference.*

This chapter should be takes as an addition and application of presented structural treatment.

The successions as dynamical phenomena are applied to *simulation the evolution of lichens communities* [16] and other processes.

Conclusion

Is based on the fact that structure is presentable in the form of a graph and the isomorphic graphs have the same structure. It must be recognise that the essence of structure appear in relationships between its elements and belonging of these to the positions.

We can assert that the structure of graph is formerly no explored (studied) and the meaning of the structure has changed to an undetermined adjective. Here is demonstrated the necessity of structure model in research the graph structure. As well the importance of the positions in structure, and the role of position- and sign structures in the investigation of the structures. Also mutual treatment the structure and it complement. On the other hand is structure an inseparable attribute of all the really existing objects.

In central place here is the problem of *recognition the binary positions* with deep identification of the element pairs. It works and contra examples do not find. The essence of structure and all its essential characteristics are based on the binary signs and binary positions. Also the essence of notorious reconstruction problem and the systems of adjacent structures are reduced to the positions. Such approaching mode can be also *a mathematical problem* what need to prove or to disprove.

The *time complexity* of forming the structure model depends only from the number of elements and is polynomial.

It is drawn attention to the fact that, based on the model of structure, are united and mutually developed different traditional graph theoretical problems, such as the *orbits detecting (transitivity domains of automorphisms, positions), multiplication of adjacency matrices, canonical representation of graphs, isomorphism, reconstructions of graphs and enumeration of graphs.*

On the basis of structurally treated traditional problems opens the model of structure also such "hidden sides" of graphs, as *clique-regularity, girth-regularity, relations between regularity and symmetry properties, sign structures, position structures, the coexistence of structural transformation and reconstruction, systems of the adjacent structures and their various probabilistic characteristics, using its for real objects and other.* Unfortunately, not all the examples do not finds room here. On the theoretical foundations are described in beginning of publication [36] and on the practical examples are presented in the publications [35, 39, 40], that are available also in digital form.

In principle, structure model can be considered as a development of adjacency matrix to a „relationships matrix". To the first attempt to this direction can be considers the „distance matrix", presented already in year 1973 [42].

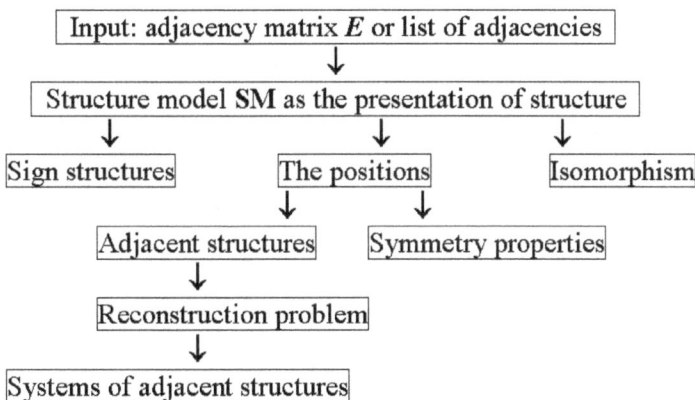

Such approach is called also as *semiotics of structure* and can be treated as *using of graphs for explanation of the meaning of structure.*

It must be hoped that this writing will help to better understand the nature of the structure of the graph.

References

1. Babai, L. On the isomorphism problem. *Unpublished manuscript, 1977.*

2. Babai, L. Luks, E. Canonical labelling of graphs. – *Proc. 15th ACM Symposium on Theory Computing, 1983, 171-183.*

3. Berge, C. Graphs and Hypergraphs. *Paris, Dunod, 1970*

4. Cayley, A. On the theory of the analytical forms called trees. – *Phil. Mag. (4) 13 (1857), 172-176.*

5. Collatz, L., Sinagowitz, U. Spektren endlicher Graphen. –*Abh. Math. Sem. Univ. Hamburg, 21 (1957), 63-77.*

6. Dharwadker, A., Tevet, J.-T. (2009) The Graph Isomorphism Algorithm. ISBN 9781466394377. *Amazon Books, 2009.*

7. Eilenberg, S., Mac Lane, S. Relations between homology and homotopy groups of spaces. – *Annals of Mathematics 51 (1950), 514-533.*

8. Futuyama, D. J. B. Evolutional Biology. *Sinauer Associate, Sunderland, Massauchusetti, 1978. ISBN 0-87893-199-6.*

9. Gurevich, Y. From Invariants to Canonization. – *The Bull. of Euto. Assoc. for Comp. Sci. No 63, 1997.*

10. Harary, F. Graph Theory. *Addison-Wesley, 1969.*

11. Harary, F., Palmer, E. M. Graph Enumeration. *Academic Press, 1973.*

12. Hermes, H. Semiotik: eine Theorie der Zeichengestalten als Grundlage für Untersuchungen von formalisierte Sprachen. *Leipzig: Hirzel, 1938.*

13. Hoffman, C. Group-Theoretic Algorithms and Graph Isomorphism. *Springer, 1982.*

14. Kalman, R. E. Topics in Mathematical System Theory. *N. Y. 1969.*

15. Locke, S. www.math.fau.edu/locke/isotest.

16. Martin, J., Tevet, J.-T. On the interrelations between structure, dynamics and evolution of lichen synusiae. – *Proc. Estonian Acad. Sci., Biol., 37 (1988) No 1, 56-66.*

17. Mathon, R. Sample graphs for isomorphism testing. – *Proc. 9th S-E. Conf. Combinatorics, Graph Theory and Computing, 1980, 499-517.*

18. Mayer. J. Developments recents de la theorie des graphes. – *Historia Mathematica 3 (1976) 55-62.*

19. Nechepurenko, M et al. *М. Нечепуренко и др. Алгоритмы и программы решение задач для графов и сетей. Новосибирск, 1990.*

20. Новая философская энциклопедия, *Москва. 2001.*

21. Nöth, W. Handbook of Semiotics. *1995.*

22. Podsiadlo, B. http://web.me.com/blazej.podsiadlo/poudis/Graph_Isomorphism.html

23. Praust, V. Tevet's algorithm for graphs recognition. *Tallinn, 1995.*

24. Read, R. C., Wilson, R. J. An Atlas of Graphs. *Oxford, 1998.*

25. Schmidt, H. Philosophisches Wörterbuch. *Stuttgard, 1991.*

26. Strongly regular graphs I. http://poeple.csse.uwa.edu.au/gordon/remote/srgs

27. Strongly Regular Graphs II. http://mathworld.wolfram.com/StronglyRegularGraph

28. Tevet, J.-T. (1984) The symmetry phenomenon on the aspect of structural principle (*in Estonian*). – *Schola Biotheoretica X, Tartu, 1984, 84-93.*

29.   (1990) Interpretations on some Graph Theoretical Problems. *Edition of Estonian Acad. Sci., Tallinn, 1990.*

30.   *(1999) Appendix to Structure Semiotics: A System of Graphs, their Characteristics and Changes.* S.E.R.R., Tallinn, 1999, 95 pp.

31.   *(2001) Semiotic Testing of the Graphs. Principles, Using, Developments.* S.E.R.R., Tallinn, 2001.

32.   *(2002) Isomorphism and Reconstructions of the graphs. A Constructive Approach and Development.* S.E.R.R., Tallinn, 2002.

33.   *(2004) Heuristic Algorithms for Structure Processing of the Graphs. S.E.R.R., Tallinn, 2004.*

34.   *(2006) Structure-semiotic approach to the graphs. S.E.R.R., Tallinn, 2006.*

35.   *(2008) Constructive presentation of the graphs: a selection of examples. S.E.R.R., Tallinn, 2008.*

36.   *(2010) Hidden sides of the graphs. ISBN 9789949213108. S.E.R.R., Tallinn, 2010.*

37.   *(2012a) Semiotic Modelling of the Graphs. ISBN 9789949302475, S.E.R.R., Tallinn, 2012.*

38.   *(2012b) The story of SERR. ISBN 9789949308774, S.E.R.R., Tallinn, 2012.*

39.   *(2013) Nature of the Structure. ISBN 9789949334650, S.E.R.R., Tallinn, 2013.*

40.   *(2014) Some examples and explanations about the structure of graphs. ISBN 9789949336333. S.E.R.R., Tallinn,2014.*

41. Titov,V. *Титов, В.* О симметрии в графах. – *Вопросы кибернетики, 15, N2, 1975, 76-109.*

42 Toida, S. Isomorphism of graphs. – *Proc. 16th Midwest Symp. Circuit Theory, Waterloo, 1973, XVI5.1-5.7.*

43. Tutte, W. T. Graph Theory As I Have Known It. *Clarendon Press, Oxford, 1998.*

44. Ulam, S. M. A Collection of Mathematical Problems. *Wiley, New York, 1960.*

45. Weisfeiler, B. On Construction and Identification of Graphs. – *Springer Lect. Notes Math., 558, 1976.*

46. Zykov, A. *Зыков, А.* Основы теории графов. *«Наука», Москва, 1987.*

www.ingramcontent.com/pod-product-compliance
Lightning Source LLC
Chambersburg PA
CBHW040743200526
45159CB00023B/1606